806

39
40
71

Atmospheric Pollution

For my family,
especially Elizabeth, Sally and Clare

Atmospheric Pollution

Causes, Effects and Control Policies

Derek Elsom

Basil Blackwell

Burgess
TD
883
.E 46
1987g

c. 1

First published 1987
Reprinted 1989

Basil Blackwell Ltd   /
108 Cowley Road, Oxford, OX4 1JF, UK

Basil Blackwell Inc.
432 Park Avenue South, Suite 1503
New York, NY 10016, USA

British Library Cataloguing in Publication Data
Elsom, Derek M.
    Atmospheric pollution: causes, effects and control policies
    1. Air – Pollution
    I. Title
    363.7′392   TD883

    ISBN 0-631-13815-3
    ISBN 0-631-15674-7 Pbk

Library of Congress Cataloging in Publication Data
Elsom, Derek M.
    Atmospheric pollution.

    Bibliography: p.
    Includes index.
    1. Air – Pollution – Environmental aspects.
2. Pollutants – Environmental aspects.   3. Environmental policy.   I. Title
TD883.E46 1987   363.7′392   87-11645
ISBN 0-631-13815-3
ISBN 0-631-15674-7 (pbk.)

Typeset in 10 on 11½pt Sabon by
Dobbie Typesetting Service, Plymouth, Devon
Printed in Great Britain by
T. J. Press (Padstow) Ltd

Contents

Preface

Atmospheric pollution is a major problem facing all nations of the world. Rapid urban and industrial growth has resulted in vast quantities of potentially harmful waste products being released into the atmosphere. Societies have been reluctant to accept, or have simply failed to recognize, the limitations of the cleansing properties of the atmosphere. The consequence has been that air pollution has affected the health and well-being of people, has caused widespread damage to vegetation, crops, wildlife, materials, buildings and climate, and has resulted in depletion of the scarce natural resources needed for long-term economic development.

The worst air pollution has occurred in and around urban–industrial areas. The seriousness of atmospheric pollution for urban communities, as shown by the effects of short-term pollution episodes or accidental releases of large quantities of a pollutant, has led to the introduction of national pollution-control policies which have largely been aimed at tackling local pollution problems. Initially, these national policies gave little or no consideration to the phenomenon of exported or imported pollution. Inevitably, the transport of pollutants over long distances has created international and global pollution problems, and problems such as acid rain, ozone episodes, and accidental releases of toxic chemicals and ionizing radiation have emerged. Areas distant from sources of pollution, once perceived as being too remote to be affected, now experience acid rain which threatens irreversible damage to sensitive aquatic and terrestrial ecosystems. Global-scale problems include increasing the atmospheric concentration of carbon dioxide which may be causing a change in the world's climate, and chlorofluorocarbons (CFCs) which may be depleting stratospheric ozone. Today nations are at last beginning both to recognize the seriousness of transfrontier and global pollution problems and to realize that solutions will emerge only when nations co-operate together on effective regulatory action.

The scope of the atmospheric pollution problem and the nature, sources and effects of pollutants are introduced in chapter 1. A wide range of pollutants are then surveyed in the following five chapters and emphasis is placed on assessing the effects of the pollutants on the health of communities, animals and vegetation, materials, and the atmosphere (e.g. weather and climate). Chapter 7 explains the alternative strategies which may be adopted to tackle polution problems. The development and effectiveness of national pollution control policies are examined in chapters 8 to 10 including those of the United States, the United Kingdom (and the European Community), the Soviet Union and the People's Republic of China. Differences and similarities in pollution problems and control policies between socialist and capitalist states, and between developing and developed countries, are highlighted in these chapters. Finally, the progress towards international collaboration on such problems as acid rain, stratospheric-ozone depletion and the build-up in global concentration of carbon dioxide is examined, and future prospects assessed, in chapter 11.

The book is written for undergraduate students studying the atmospheric pollution problem within the disciplines of environmental studies, geography, environmental planning, environmental engineering, environmental biology and the broader social sciences. The structure of the text has been developed from the experience of teaching a course on atmospheric pollution as part of the Modular Degree Course at Oxford Polytechnic. This system permits students from a wide range of disciplines to select a few courses from outside their immediate discipline, which means that the course leader must develop an approach to studying the subject matter which is suitable for students with varying backgrounds. In this context, I am greatly indebted to the students at Oxford who have unknowingly played a crucial part in developing the present text.

My own interest in atmospheric pollution was first stimulated whilst taking the geography degree and then the MSc course in Applied Meteorology and Climatology at the University of Birmingham. This was followed by undertaking research in air pollution meteorology at University College London and the University of Manchester under Professor Tony J. Chandler. At that time, Professor Chandler was closely involved in atmospheric pollution studies for the United Kingdom Royal Commission on Environmental Pollution and the World Meteorological Organization, and I regard myself as very fortunate in having had the benefit of his guidance.

The preparation and writing of this text has been helped by the stimulus provided by many people, including Ron Barnes of the Esso Research Centre; Alistair Keddie, former head of the Air Pollution Section at Warren Spring Laboratory; and my colleagues in the Geography Section at Oxford Polytechnic, especially John Gold, Martin Haigh, Alan Jenkins, Heather Jones, Peter Keene, Anna Kilmartin, David Pepper and Stephen Ward. A visit to the People's Republic of China in October 1985 at the invitation of Professor Qu Geping, Director of their National Environmental

Protection Agency, provided an invaluable opportunity to gain an insight into that country's pollution problems and approach to pollution control. My thanks are extended to numerous unnamed individuals from many other organizations including the United States Environmental Protection Agency, the United States South Coast Air Quality Management District, and the United Nations Environmental Programme based in Nairobi who have so helpfully dealt with my frequent requests for information. Finally, this study would not have been completed but for the understanding and support shown by my wife, Elizabeth, and children, Sally and Clare, especially given that so many evenings and weekends were spent with this text rather than with them.

Derek Elsom

Part I

The Nature, Sources and Effects of Atmospheric Pollution

1 The Atmospheric Pollution Problem

1.1 Defining Atmospheric Pollution

There are many different views as to what constitutes pollution of the atmosphere. To some people, pollution implies the increase, or even decrease, of any atmospheric constituent from the value that would have existed without human activity. Given that our planet's atmosphere has undergone profound changes in its constitution throughout its lifetime and that volcanic eruptions, forest fires and sand storms cause marked local and regional variations in atmospheric constituent values, then such a definition is of limited use. Other definitions of atmospheric pollution are more useful, such as 'the presence of substances in the ambient atmosphere, resulting from the activity of man or from natural processes, causing adverse effects to man and the environment' (Weber, 1982). An expanded version of this definition will be employed in this text – namely, that air pollution is defined as 'the presence in the atmosphere of substances or energy in such quantities and of such duration liable to cause harm to human, plant, or animal life, or damage to human-made materials and structures, or changes in the weather and climate, or interference with the comfortable enjoyment of life or property or other human activities'.

1.2 Types of Pollutants

Until well into this century, air pollution was for most people synonymous with suspended particulate matter (soot, smoke) and sulphur dioxide. These are waste products produced mainly by domestic heating equipment, a wide range of industrial plants, and power plants. As the twentieth century has progressed, concern for pollution of the atmosphere has ranged across a large number of pollutants. The tremendous increase in the use of petroleum products, particularly in petrol-powered motor vehicles, introduced several new pollutants (World Health Organization, 1972). Exhaust emissions of oxides of nitrogen, carbon monoxide, hydrocarbons,

Figure 1.1 The extensive damage to central European forests, attributed to acid rain, has increased the urgency for international collaboration on pollution-control measures
Source: R. A. Barnes

and lead added greatly to the pollution of urban areas. From the emissions of oxides of nitrogen and hydrocarbons are also produced the secondary pollutants of photochemical oxidants. Waste energy in the form of noise and heat also contribute to the list of local pollutants. The development of new industries introduced the problem of toxic chemicals, while nuclear-power production and atomic-weapons testing highlighted ionizing radiation as a pollutant.

Until the 1960s, pollutants were generally only regarded as a problem in the vicinity of individual emission sources or within or near urban areas. Subsequently, studies showed that pollutants were being transported over long distances and were causing adverse effects on the environment at locations far removed from the source of the emissions. Long-range transport of sulphur and nitrogen compounds across national boundaries resulted in increased acidity of precipitation (acid rain) at distant locations and created an international pollution problem (figure 1.1). Improved monitoring and detection of trends in atmospheric constituents, together with the modelling of atmospheric processes, have highlighted the fact that pollution problems can become global in extent. Pollutants of current local or regional concern, such as anthropogenic heat and toxic chemicals, have been projected to become global pollution problems of the future. The increase in atmospheric concentration of carbon dioxide has been claimed both to have the potential to alter the Earth's climate and to have profound effects on world agriculture. Stratospheric ozone depletion by a variety of anthropogenic compounds is another global pollution issue which emerged in the 1970s.

1.3   The Unfairness of Pollution: Pollution not Borne by the Polluters

It should be remembered that to some extent pollution is economically beneficial to some people. When a factory discharges wastes into the atmosphere, it is presumably adopting the cheapest way of disposing of its unwanted materials. The act of polluting the air therefore keeps the price of its products lower than would otherwise be the case. Consequently, the factory owners sell more products and make greater profits, and the consumers of the products buy them at a lower price than they would have to pay if pollution of the atmosphere were avoided. The pollution situation thus exists because of society's desire for consumer products and its eagerness to purchase them at the lowest short-term market cost, even if this means generating substantial quantities of waste materials.

Although the atmospheric pollution may result in adverse effects on the nearby population (including the factory work-force and the factory owners) which is enjoying the benefit of cheaper products, it also affects people (and the environment of future generations) who may have derived no benefit whatsoever from the polluting factory. In this situation, those who are making use of the atmosphere for waste disposal are not paying for its use but are passing the costs on to society in general. It may even be that those who suffer the consequences of pollution (e.g. the effects of acid rain) do not even reside in the same country as those who do the polluting, while for some pollutants, such as toxic chemicals and radioactive wastes, those who may suffer most are as yet unborn. When polluters are forced to pay for pollution control (referred to as the 'Polluter Pays Principle'), the costs previously borne by outsiders are internalized in the production process. The expense of pollution control is then passed on to the consumer in higher prices for products or services, effectively transferring to the consumer the real costs that were previously borne by society in the form of ill-health, property damage and environmental damage.

1.4   The Need to Reduce Pollution

The act of polluting the atmosphere may be financially beneficial to some members of society but for society in general, especially in the long term, pollution is very costly. Illness or premature death due wholly or in part to air pollution places a great burden upon society by way of increased costs of medical treatment and through the loss of manpower. In addition, air pollution adversely affects soils, water, crops, vegetation, human-made materials, buildings, animals, wildlife, weather, climate, and transport-ation, as well as reducing economic values, personal comfort and well-being (Purdom, 1980).

Given that the reduction of pollution is itself costly, it is necessary for society to decide how much of its limited resources it should allocate for pollution control. Reducing pollution to the point where the costs of doing so are covered by the benefits from the reduction in pollution would seem the obvious first step. However, in practice, the assessment of costs and benefits is very difficult: one has to weigh up short-term versus long-term costs and benefits within the geographical area (national or global) under consideration and the effects of a changing level of technology and fluctuating resources, as well as taking into account social and political value-judgements which change over time. Ultimately, because of such complexities, the degree of pollution control adopted becomes a social and political decision.

1.5   The Environmental Movement in Developed Countries

In the late 1960s, environmental deterioration emerged as a political issue more or less simultaneously in several industrialized countries (figure 1.2).

Prior to this, public concern for environmental quality had generally been limited to local pollution problems. This quickly changed as citizens became aware that environmental problems were becoming both more numerous and more serious: the Environmental Movement was underway. By the end of the decade, national and international dimensions of the pollution problem were being raised by pressure groups and the media. Helped by the media, citizens in the developed countries increasingly demanded that government action be taken both to arrest further deterioration of the environment and to improve its quality (Jacobson and Kay, 1983). As a consequence, many countries introduced environmental legislation and created environmental protection agencies, and in addition, international organizations such as the United Nations turned their attention to environmental protection.

During the early 1970s, public and media concern for pollution problems decreased in most industrialized countries (figure 1.2). As the developed countries moved from planning to the implementation of environmental protection, the costs of such action became clearer and a lengthy debate developed concerning the extent to which these costs were worth paying and who should pay them. In addition, environmental problems that had seemed simple to solve when first examined proved more intractable when protection efforts were actually attempted. In several instances, greater understanding of complex problems was required before effective action could be planned. Furthermore, the embargo and price increase imposed

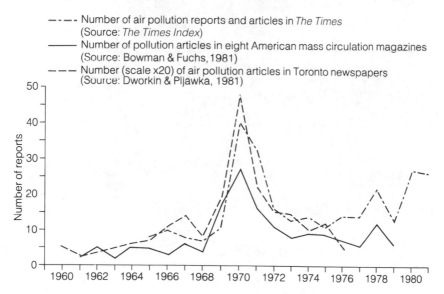

Figure 1.2 The coverage of pollution issues in newspapers and magazines highlights the almost simultaneous rise in international concern for pollution in the late 1960s

Figure 1.3 The Chernobyl nuclear power station: the accidental release of radionuclides in 1986 resulted in about 30 deaths in the first few weeks, but may result in tens of thousands of deaths in the Soviet Union and Europe in the following decades
Source: Camera Press (TASS) London

by the member states of the Organization of Petroleum Exporting Countries (OPEC) in 1973 resulted in an energy crisis, and as a consequence, public and government concern for energy problems exceeded that for environmental problems. Many developed countries began to face economic difficulties in the 1970s and this also reduced enthusiasm for environmental protection activities. Nevertheless, in the 1970s, public concern and political awareness of pollution issues remained at a higher level than during the early and mid-1960s.

In the 1980s, public support for measures to improve the environment remains high and continues to grow despite slower economic growth and unemployment. This interest stems from a number of factors, including higher incomes, a greater awareness of environmental needs and growing recognition of the benefits of environmental protection measures. Acid rain, lead, toxic chemicals and ionizing radiation are some of the principal pollutants which have been the foci of increased concern and attention. Accidents involving ionizing radiation at Three Mile Island, USA, in 1979, and at Chernobyl, in the Soviet Union, in 1986, dramatically heightened public interest in matters of atmospheric pollution (figure 1.3).

1.6   Pollution Control in Less-developed Countries

Even though pollution was almost as serious in some of the rapidly expanding cities of the less-developed countries – Mexico City, São Paulo, Lagos, Cairo, Beijing and Bangkok – as in Los Angeles, London, or Tokyo, nowhere in the less-developed countries was pollution a public issue in the same way that it was in the developed countries. In the late 1960s the policies of the less-developed countries were focused on issues of economic development, and to the limited extent that the public were involved in policy-making, they shared this commitment. It is true that some policies, such as those designed to improve the quality of drinking water and to provide sanitation facilities, involved environmental issues, but these were seen as as an aspect of economic development rather than as efforts to protect the environment (Jacobson and Kay, 1983).

When the question of international action to protect the environment was first raised, many less-developed countries expressed concern that environmental protection measures would slow down their rates of economic and industrial growth. It was pointed out that industrialization had satisfied many of the needs of developed countries and such countries could now afford to express their concern about pollution, whereas less-developed countries could not yet afford to do so. When a higher standard of living had been achieved, then, they argued, would be the time to attend to environmental matters. One country which had followed this line of reasoning was Japan which, in the late 1940s and 1950s, had adopted a policy of economic growth at all costs but by the 1970s regarded pollution as a social crime.

The introduction of environmental protection legislation in developed countries in the early 1970s even created economic advantages for some less-developed countries. Thus, for example, many new industrial plants manufacturing toxic chemicals were established in these countries because their less strict environmental standards meant less expenditure on pollution-control equipment and fewer delays in getting the necessary authorization for plant construction. However, the 1972 United Nations Conference on the Human Environment at Stockholm and the resulting United Nations Environment Programme (UNEP) helped many less-developed countries to begin to recognize the importance to their own countries of protecting the environment, since environmental deterioration would lead to the loss of resources on which people would depend in the long term. They came to realize that concern for the environment need not block their countries' progress toward industrialization, but rather, could actually be built into economic development programmes. In addition, incidents such as the tragedy at Bhopal, India, in December 1984, in which 2500 people died when methyl isocyanate was accidentally released from the Union Carbide chemical plant, has considerably offset the economic arguments for allowing lax pollution-control requirements in such industrial plants.

Many current pollution problems are a consequence of the rapid increase in global population and the clustering of that population in large urban conglomerations in which waste-producing industries are concentrated (40 per cent of the world's population lives in urban areas). This suggests that the less-developed countries will become more significant polluters in the future as their rate of population growth increases while that of developed countries slows down. In 1950 there was only one city (Buenos Aires) in the developing world with a population of more than four million; by 1975 there were 17 (compared with 13 in developed countries), and by 1980 the total had risen to 22 (compared with 16). By the year 2000, there are likely to exist some 61 cities of over four million inhabitants in the developing countries, as against 25 in developed nations, and the total urban population in the developing world is predicted to double between 1980 and the end of the century (Holdgate et al., 1982b).

Since the 1970s, although economic and industrial growth remains a priority in less-developed countries, they have undertaken some reassessment of the need for pollution control measures. With developed countries moderating for various reasons their original enthusiasm for environmental protection, the countries of the world have moved towards a consensus on action to protect the environment. Although the debate between the developed and less-developed countries continues to reflect each side's different problems and priorities, the distance between the two sides has considerably diminished and progress is being made on pollution control throughout the world.

1.7   Cure is No Substitute for Prevention

Economic and technological achievements over the last generation have been considerable: they have brought immense and worthwhile benefits to people throughout the world. However, the benefits have often been bought at the cost of a deterioration in environmental quality (UK Royal Commission on Environmental Pollution, 1971). Industrialization transforms natural resources and in so doing produces waste materials and waste energy which require disposal (figure 1.4). Industrial activity is a necessity, but society needs to reconsider its priorities and it needs to place greater emphasis on technological developments that are materials-saving, since the latter reduce the quantity of new materials used and consequently reduce the wastes disposed of in the environment. Technological changes can improve utilization of wastes such that materials that were formerly wastes may become useful, thereby no longer needing to be discarded. With high-grade raw materials becoming increasingly exhausted, a sensible materials policy is needed which emphasizes the repair, reuse, and recycling of goods which are currently discarded.

However, technological innovation alone will not solve the world's pollution problems: for example, an acute air-pollution problem may be solved by converting the air pollutants into a solid or sludge, but this technological fix then results in contamination of the land or water and, in the long term, such wastes may eventually reach the atmosphere anyway. This points to the need for pollution problems to be tackled ultimately in an holistic and preventive manner. An holistic approach to pollution problems requires that the 'best available environmental option' is adopted for each waste disposal problem such that the disposal method which is best for the environment is selected, having taken into account technical and economic considerations. Even so, in the long term, the only way to reduce the pollution problem is to reduce generation of all wastes. Energy, transportation, industry, agriculture and land-use management policies need to give much greater consideration to anticipating and preventing pollution problems. Curative pollution-control policies are no substitute for preventive policies (OECD, 1980a).

1.8   Understanding Atmospheric Pollution Problems

A sound understanding of both the causes of atmospheric-pollution problems and the nature and effects of atmospheric pollutants is needed

Figure 1.4 Industrial complexes such as this concentration of petrochemical plants at Grangemouth, Scotland, introduce a wide range of waste products into the atmosphere

before appropriate and effective pollution control can be introduced, and such a framework is provided in chapters 2–6, which examine each major atmospheric pollutant or group of pollutants in turn.

1.9 Formulating the Levels of Human and Environmental Protection against Pollution

Before effective air pollution control strategies can be developed, clear objectives or policy goals need to be established. Full protection of the health and welfare of people and the environment may be the ultimate goal, but unless pollutants are entirely eliminated from use, complete protection is unattainable. Consequently, society must decide on the level of health risk and environmental protection that is acceptable. This decision is influenced by social, economic, technological and political factors as well as by our degree of understanding of atmospheric pollutants and their undesirable effects. Knowledge of the nature, sources, and effects of pollutants are essential if realistic goals and effective control strategies are to be employed.

Local, national and international policy goals are usually of two types, short-term and long-term. Short-term goals are concerned with seeking the reduction of pollution concentrations (or elimination for all but essential uses in the case of some toxic (poisonous) pollutants) to a 'threshold' level below which no adverse effects on public health are detectable. Long-term goals seek either an increase in the safety margin below the health-related threshold level, further banning of non-essential substances, or lowering of levels of pollution so as to reduce or eliminate adverse effects on non-human receptors (the environment) such as damage to vegetation, materials, and animals. There are broadly speaking five main types of damage caused to the environment by air pollutants:

1 damage to vegetation (including crops)
2 damage to animals, birds and insects
3 damage to human-made materials (painted surfaces, rubber, nylon, metals)
4 soiling of materials (clothing, buildings, etc.)
5 weather and climatic changes (smogs, reduced solar radiation, visibility deterioration, surface temperature increases)

The confidence with which short- and long-term goals may be specified in quantitative terms depends upon the extent and quality of research into the harmful effects of pollutants. Adequate understanding of the effects of pollutants on public health relies on evidence obtained from studies of two types: epidemiological studies and toxicological studies. Epidemiological studies are concerned with the effects occurring in human

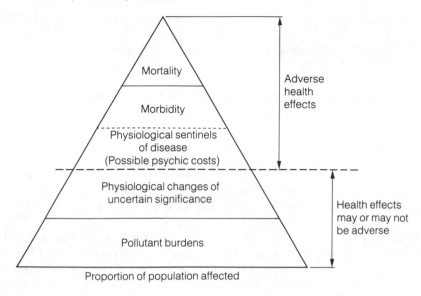

Figure 1.5 Health effects of air pollution
Source: Shy and Finklea, 1973

communities exposed under natural conditions (Goldsmith, 1986). Laboratory studies on people and animals, in which the level, duration, and conditions of exposure are under the control of the investigator, are included in the term 'toxicological' (World Health Organization, 1972). Both types of study attempt to determine the nature and severity of the health effects on the population.

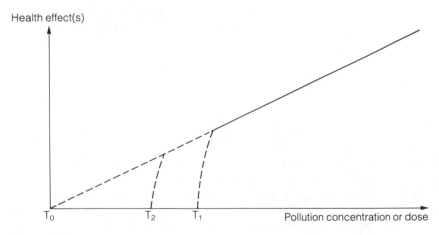

Figure 1.6 Dose-response relationships between pollution and health effects

The health effects of pollutants cover a wide spectrum of biological response (figure 1.5): the more severe effects such as death and chronic illness will be manifested in a relatively small proportion of the population, while many more individuals in an exposed community will respond with altered physiology or pollutant burdens (pollutants accumulating in the body). It is important to establish the 'dose-response relationship' for a pollutant or for a combination of pollutants, should additive or synergistic effects occur. For a particular pollutant and specific health effect, a threshold value may be recognized which corresponds to a pollution concentration or dose below which no health effect is detectable (threshold T_1 in figure 1.6). However, fresh evidence or more sensitive measurements may result in the lowering of a threshold (to T_2). In cases where any infinitesimal exposure causes a reaction among some individuals in the population, one may conclude that the threshold value is zero (T_0).

Epidemiological and toxicological evidence is complementary in that both types of evidence are desirable to form an acceptable judgement about the health effects of pollutants and to produce dose-response relationships. While epidemiological studies are relevant because they are undertaken in the natural setting, many complicating factors are present, some of which only a well-controlled laboratory experiment can eliminate. Furthermore, epidemiological evidence may be too limited in the range of levels of pollution exposures studied to be able to establish clearly the mathematical form of the dose-response relationship. In such a situation, laboratory studies of animals may be essential in showing the way in which, say, the frequency of induced cancers is related to a dose of ionizing radiation or some other carcinogen (substance that causes cancer). Nevertheless, many complicating factors may be present in both epidemiological and toxicological studies, and one should be aware of this when assessing the validity and wider applicability of the research results.

In both epidemiological and toxicological studies, the health responses of the sampled population depend upon the sensitivity of the individuals forming that population. A number of factors affect the sensitivity of the population, including age, sex, general state of health and nutrition, concurrent exposure, pre-existing diseases, temperature and humidity, and the level of activity at the time of exposure. Thus, for example, increased activity leads to a greater ventilation rate, increasing the overall intake and penetration of some pollutants and thereby enhancing their effects (Lawther et al., 1975). In general, the elderly, the very young, those in poor health, cigarette smokers, the occupationally exposed, and those with pre-existing chronic bronchitis, coronary heart disease and asthma are more vulnerable to exposure to pollutants. Given the problems of matching populations for important variables such as smoking, occupational and previous medical history, many epidemiological studies decide to study school children in order to minimize differences. Nevertheless, such studies are faced with enormous problems in attempting to overcome variables such as housing conditions, social class and indoor pollution exposure (Pengelly et al., 1984).

Epidemiological studies focus on the health effects of ambient (outdoor) pollution exposures and often ignore indoor exposures. Yet it should be remembered that urban populations spend proportionately more of their time inside buildings (and vehicles) than out of doors. Occupational indoor exposure may for many individuals contribute more to their pollutant loading than outdoor pollution levels. In their homes, people may be subject to a wide variety of pollutants such as carbon monoxide, oxides of nitrogen and suspended particulates released from cigarettes and maladjusted cooking and heating appliances, formaldehyde released from urea-formaldehyde foam-insulation and plastics, radon emitted from radium-containing building materials, asbestos fibres from building materials, and chlordane and pentachlorophenol (PCP or penta) in homes built from or employing treated wood (Walsh et al., 1983). Children in particular appear to be susceptible to indoor pollution, with several investigators reporting increased rates of acute respiratory disease among children of smoking parents versus children of non-smokers (Binder et al., 1976; Naumann, 1973). Similarly, Melia et al. (1977) and Speizer et al. (1980) reported increased respiratory illness among school children in which gas was used for cooking, compared with children from homes where electricity was used. This 'cooking effect' was probably caused by high indoor levels of nitrogen dioxide produced by the gas stoves (Melia et al., 1978). Internal pollution is likely to increase because of our increasing emphasis on energy conservation, controlling draughts, and using high proportions of recirculating air. Although recirculation systems may include filters to trap particles, gas contamination is not removed and is therefore likely to build up. Ideally, studies of the health effects of pollutants should be based upon pollution measurements derived from personal samplers. However, they are not in widespread use, except in specific occupational health studies, and are therefore inappropriate for large-scale epidemiological studies of the general population.

Analyses of the relationships between community health and pollution levels require that the ambient pollution concentrations recorded by monitors are representative of the pollution exposure of residents. However, the validity of this assumption is questioned when one considers that the population is mobile and spends varying lengths of time in different locations, thereby changing their pollution exposure each time. Even for residents who both live and work in the same part of a city, one has to decide whether one monitor is adequate to provide pollution measurements representative of that part of the city. The area over which pollution measurements from a single monitor are applicable will depend upon the distribution of emission sources, the pervasiveness of the pollutant, topography, meteorological and synoptic conditions, and many other factors (Elsom, 1978a, 1979; Munn, 1981). In addition, some monitors may be sited on the roof of a building either for protection from vandalism or for convenience, which means they may not provide representative measurements of the pollution at the height at which people breathe.

If one is satisfied that monitored pollution levels are representative of an individual's or a community's pollution exposure, the problem still remains of deciding what pollution sampling period one should employ in the analyses. The effect of a brief exposure to a very high concentration (acute effect) can not often be separated from a situation of a steady exposure to a low level of pollution (chronic effect). Various time-periods may have to be selected for study to allow consideration of annual, seasonal, daily, hourly, or peak pollution-concentrations. Fortuitously, for a wide range of averaging times, pollutants and localities, frequency distributions follow a log-normal distribution, and this explains why the geometric mean, equivalent to the median in such a distribution, is regarded as more appropriate than the usually slightly higher arithmetic mean (Pollack, 1975).

When applying the results of dose-response relationships derived from a study in one community to another, one must recognize not only that population sensitivity varies but that the pollutant also varies. Thus, for example, suspended particulate matter in an urban area is derived from diverse sources and types of activity such that the precise nature of those particulates is unique to a particular city and even to a particular time. For epidemiological studies this consideration is made even more important when it is realized that the method of measurement of suspended particulates may vary. Whereas in Europe, measurements of suspended particulates are based on soiling properties, in the United States and elsewhere the monitoring technique is based on weight. The former provides an index of smoke (or soot) in the atmosphere, associated with the incomplete combustion of fuel, while the latter measurement is a wider concept that includes all particulate material which by virtue of its particle size remains in suspension for long periods. Even within countries, different measurement techniques may be employed: for example, most of the early epidemiological studies concerning photochemical oxidants were undertaken in the Los Angeles area with oxidants being measured by the unbuffered potassium iodide method. This method yields oxidant values 15–25 per cent lower than values obtained with the 1 or 2 per cent neutral-buffered potassium iodide method subsequently employed around the world (World Health Organization, 1979b).

Given the existence of so many complicating factors and study limitations, it is not surprising that health and environmental effects of pollutants are difficult to quantify and thus are the subject of considerable debate and controversy. Nevertheless, for some pollutants a consensus of agreement does exist concerning the general levels at which specified adverse effects are detectable. Valuable independent assessments of diverse research studies are provided by international organizations concerned with the problem of environmental pollution, such as the World Health Organization and the International Commission on Radiological Protection. For many countries, the short- and long-term goals published by these organizations are readily adopted as national goals. In the remaining chapters of Part I, which examine the principal atmospheric pollutants to which the general population is exposed, emphasis is given to independent assessments of the health and environmental effects of pollutants as provided by the international organizations.

2 Particulates and Gaseous Pollutants

2.1 Suspended Particulate Matter and Sulphur Dioxide

The term 'suspended particulate matter' refers to the wide range of finely divided solids or liquids dispersed into the air from combustion processes (heating and power generation), industrial activities, and natural sources. Suspended particulates range in size from 0.1 up to about 25 μm in diameter.[1] The constituents of suspended particulate matter vary over time and space, although typical constituents in urban areas include carbon or higher hydrocarbons formed by incomplete combustion of hydrocarbon fuels. Up to 20 per cent of the total suspended matter may consist of sulphuric acid and other sulphates (as much as 80 per cent of the particles less than 1 μm in diameter).

A major problem studying the health effects of suspended particulate matter is that there are two differing techniques widely employed to monitor this pollutant. Whereas in Europe, measurements of suspended particulate matter are based on soiling properties,[2] in the USA the

---

[1] One micrometre (1 μm) equals 10^{-6} metre. The term micron (μ) is the non-SI name given to the micrometre.

[2] The smoke-shade or reflectance method draws air though a filter-paper such that smoke particles suspended in the air are retained on the paper, forming a stain. 'Smoke' is considered to include particles of approximately 10 μm diameter or less. The density of the stain depends partly on the mass of smoke particles collected and partly on the nature of the smoke. For example, the blackness per unit mass of diesel particulate matter is three times that of smoke from bituminous coal combustion and seven times that from petrol engine exhaust. The concentration of smoke in the atmosphere can be estimated by drawing a known volume of air through a filter-paper and measuring the density of the resulting stain with a photoelectric reflectometer. Usually about 2 cubic metres of air are sampled per day. A calibration curve relating the density of the filter stain to the weight of smoke particles deposited on the filter-paper has been established for 'standard urban smoke'. Thus the concentration of smoke per unit volume of air can be calculated in terms of the 'standard smoke' equivalent.

Figure 2.1 Mid-afternoon in London during a pollution episode (smog) on 30 November 1982. Ambient concentrations of smoke and sulphur dioxide were a fraction of what they were during the smogs of the 1950s and 1960s
Source: London Scientific Services

monitoring technique is based on weight.[3] The World Health Organization (1976a) recommends the former suspended particulate matter sample to be referred to as 'smoke' (or sometimes 'soot'), and the latter as 'total suspended particulate' (TSP). Although comparative evaluations of the two methods have been undertaken at a number of sites, there is no generally applicable conversion factor (Ball and Hume, 1977; Commins and Waller, 1967; Lee et al., 1972; Pashel and Egner, 1981). In general, annual average total suspended particulate concentrations are approximately 100 $\mu g/m^3$ higher than smoke concentrations. Individual daily concentrations vary greatly such that at low pollution levels the difference is large, whereas at high concentrations (of the order of 500 $\mu g/m^3$ or more) the difference is small.

Sulphur dioxide is a colourless gas emitted from similar sources as suspended particulates, especially from the combustion of coal and oil. It can react catalytically or photochemically with other pollutants to form sulphur trioxide (rapidly hydrating to sulphuric acid), sulphuric acid and sulphates. Monitoring of sulphur compounds in the atmosphere is usually restricted to the dominant gas, sulphur dioxide, using either the acidimetric method widely adopted in Europe or the West and Gaeke (or pararosaniline) method widely employed in the United States (World Health Organization, 1976a).

Suspended particulate matter and sulphur dioxide are often regarded as the 'traditional' pollutants of urban areas. The highest levels of these pollutants occurred during the sulphurous smogs to which most large industrial cities have been subjected in the past. Although the worst smogs have now passed, some cities still experience less intense smogs (figure 2.1). The term 'smog' refers to a synthesis of smoke and fog. Smogs are caused by vast quantities of the pollutants being emitted from industry and from domestic sources (for example, coal fires, apartment incinerators) during periods when meteorological conditions fail to disperse the pollution away from the city. When winter anticyclonic conditions prevail – being characterized by calm or light winds, below-freezing temperatures and a restricted mixing depth due to a stable or inversion atmospheric lapse rate – little dispersal or dilution of pollutants occurs, and pollution concentrations may build up to high levels. Where an urban area lies within a basin or valley, the low-level intense inversion of winter anticyclones forms a 'lid' and the hills surrounding the city form the sides of a box into which massive quantities of pollutants are emitted. With the moisture added to the atmosphere by combustion processes, the availability of vast quantities of condensation nuclei in the form of suspended particulates, and the low temperatures which

---

[3]Total suspended particulate matter (TSP) is collected by means of a high-volume sampler. The sampler consists of a motor and blower enclosed in a shelter. The filter surface is arranged horizontally, facing upwards, and is protected by a roof that keeps out rain and snow and generally prevents the collection of particles larger than about 100 μm. Filters are made of glass or synthetic organic fibre. The air flow rates range from 1.1 to 1.7 cubic metres per minute. The amount of suspended particulate matter is calculated by dividing the net weight of the particulate by the total air volume sampled.

increase the relative humidity, fog formation is encouraged. The fog droplets readily dissolve sulphur dioxide to produce sulphurous acid, thereby adding to the potentially harmful nature of the smog. A polluted fog is less readily evaporated by solar radiation than a 'clean' fog, so the duration of the smog or pollution episode may be prolonged. The smog, with its socially unacceptable, economically costly, unhealthy and even dangerous high concentrations, may last several days until the anticyclone weakens or moves away from the region (Brodine, 1971; Elsom, 1978a, 1979). Some action can be taken during the smog to lessen the danger to public health: for example, during the sulphurous smog of January 1985 in the Ruhr region of West Germany, when sulphur dioxide concentrations reached 800 $\mu g/m^3$, schools were closed, the authorities appealed for children and people suffering from heart disease and asthma to stay indoors, the use of private cars was prohibited, and industrial plants and power stations were required either to reduce emissions by 30 per cent or to close down.

Although suspended particulate matter and sulphur dioxide concentrations within any part of a city are largely a function of the local emissions, the pollutants may be spread throughout the city. In urban areas the city centre is frequently a few degrees warmer than the outskirts (the heat-island effect) and this may lead to the generation of a thermal breeze circulation, analogous to a sea-breeze effect. This mesocirculation takes the form of warmer air rising in the city centre and moving out to the suburbs, while cooler air is drawn into the city centre from the outskirts and rural fringe (figure 2.2). The result is that pollutants are spread throughout the urban area (Elsom, 1978b; Landsberg, 1981 Padmanabhamurty and Hirt, 1974.)

The velocity of country breezes is a function of the heat-island intensity, atmospheric stability and the surface roughness encountered. They are strongest across the steep thermal gradient characteristic of urban margins (Chandler, 1965). Maximum velocities reach 2 to 3 m/sec. (Chandler, 1961; Findlay and Hirt, 1969; Gold, 1956; Hogstrom, 1978). Padmanabhamurty and Hirt (1974) employed the heat-island intensity as an index of the occurrence and strength of country breezes in Toronto to examine the effect of the latter on pollution transfer in the city. No effect was evident until the heat-island intensity reached approximately 2°C, apparently a threshold for the inducement of a breeze. As the heat-island intensity increased above this threshold, pollution concentration in the city increased. The country breeze, while providing some relief to the outer suburbs by bringing in cleaner rural air, transfers suburban pollution towards the city centre. The breezes are characteristically intermittent or pulsating, with a periodicity of between 1.5 and 2 hours, because the cooler air brought into the city centre by the breeze temporarily reduces the heat-island intensity (Chandler, 1961; Shreffler, 1979). Country breezes may be strengthened by katabatic winds from any surrounding hills if the urban area lies in a valley or basin. Under anticyclonic conditions, the ground surface and lower

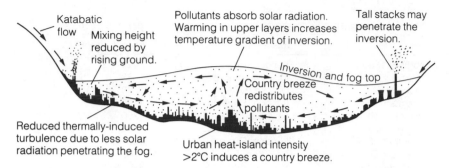

Katabatic flow

Mixing height reduced by rising ground.

Pollutants absorb solar radiation. Warming in upper layers increases temperature gradient of inversion.

Tall stacks may penetrate the inversion.

Inversion and fog top

Country breeze redistributes pollutants

Reduced thermally-induced turbulence due to less solar radiation penetrating the fog.

Urban heat-island intensity >2°C induces a country breeze.

Figure 2.2 Meteorology-pollution relationships during a smog in a valley location
Source: Modified from Elsom, 1978b

layers of the air cool rapidly at night. With cooling, the air becomes more dense and the cold air will tend to flow downhill as a katabatic wind bringing cleaner but colder air to the outskirts of the city.

2.1.1   Health Effects

The notorious sulphurous smogs which occurred in London in 1952 and 1962 and in New York in 1953, 1963 and 1966 clearly demonstrated that abrupt and substantial increases in the concentrations of suspended particulate matter and sulphur dioxide are positively associated with excess mortality (table 2.1). The individuals most susceptible were the elderly, the young, and individuals with chronic obstructive pulmonary disease and/or heart disease: for example, during or shortly after the four-day London smog of December 1952, an extra 4700 deaths occurred over and above the expected value. The increase in deaths from bronchitis was the largest single contributor to the rise in death rate, and deaths from other diseases involving impairment of respiratory functions also increased. There was an increase in the number of deaths from heart disease, which could have been due to the additional strain placed on the heart by impairment of respiratory functions or to a direct effect (World Health Organization, 1972). Peak daily concentrations of pollutants during the episode reached 6000 μg/m^3 of smoke together with nearly 4000 μg/m^3 of sulphur dioxide. Currently, with a few exceptions urban areas no longer experience such exceptional peak daily concentrations of suspended particulate matter or sulphur dioxide (table 2.2). Pollution control measures, together with socio-economic, technological and energy-source changes, have dramatically reduced the levels of these pollutants (figure 2.3).

Acute air-pollution episodes represent abrupt and unusual exposures to high pollution concentrations and produce the most obvious effects on health. Long-term or chronic exposure to moderate levels of suspended

Table 2.1 Major air pollution episodes and associated deaths up to 1966[a]

Date	Place	Excess deaths
February 1880	London, England	1000
December 1930	Meuse Valley, Belgium	63
October 1948	Donora, USA	20
December 1952	London, England	4700
November 1953	New York City, USA	250
January 1956	London, England	480
December 1957	London, England	300–800
November–December 1962	New York City, USA	46
December 1962	London, England	340–700
December 1962	Osaka, Japan	60
January–February 1963	New York City, USA	200–405
November 1966	New York City, USA	168

[a]Many pollutants, not only suspended particulates and sulphur dioxide, contributed to the excess number of deaths, such as, for example, hydrogen fluoride during the Meuse Valley and Donora episodes.
Source: Elsom, 1978b

particulate matter and sulphur dioxide also appear to impair health but the severity of this impairment is not yet clear. Aggravation and/ or causation of chronic bronchitis, asthma and pulmonary emphysema have been considered to be due to suspended particulates and sulphur dioxide. Large-scale epidemiological studies concerning both morbidity and mortality have been undertaken in London (Martin, 1964; Martin and Bradley, 1960) and in New York City (Schimmel, 1978; Schimmel

Table 2.2 Comparison of daily concentrations ($\mu g/m^3$) of smoke, total suspended particulate and sulphur dioxide in city centre commercial sites

Location	SO_2 a	SO_2 b	Smoke a	Smoke b	TSP a	TSP b
Sydney, Australia	63	246			96	246
Rio de Janeiro, Brazil	208	965	19	92		
Toronto, Canada	32	236			73	168
Bogota, Colombia	13	35	19	111		
Athens, Greece	39	165			254	438
Calcutta, India	58	105			392	751
Teheran, Iran	46	152	203	806		
Auckland, New Zealand	25	165	4	26		
Warsaw, Poland	336	2940	420	3590		
Bucharest, Romania	9	100			163	625
London, UK	125	405	26	105		
Chicago, USA	41	136			87	214
Zagreb, Yugoslavia	112	488			171	452

a = annual arithmetic mean
b = maximum daily mean
Source: World Health Organization, 1980a

and Murawski, 1976). Comparison of the results of these and other studies is complicated because of the different methods employed to sample suspended particulate matter. Generally, these studies have revealed that the adverse health effects were principally associated with suspended particulates and only to a small extent with sulphur dioxide. Mazumdar et al. (1982) reanalysed the London data of pollution and mortality data for the winters 1958–9 to 1971–2 in an attempt to differentiate more clearly between the possible separate influences of sulphur dioxide and smoke. They concluded that the pollution-mortality association was due almost entirely to smoke. Similar conclusions were reached by Mazumdar and Sussman (1983) in a study in Pittsburgh for the period 1972–7. The separate pollutant effects on morbidity have yet to be considered. The need to separate the effects of each pollutant, and especially to re-examine the health effects of sulphur dioxide, has led to increased toxicological research. Controlled laboratory studies of the effects of sulphur dioxide have shown that the adverse effects of this pollutant increase significantly when subjects undergo exercise, because switching from nose- to mouth-breathing increases the dose of sulphur dioxide to the lower respiratory tract (Anderson et al., 1974; Lawther et al., 1975). Asthmatic patients and hay fever sufferers are revealed as being particularly sensitive to sulphur dioxide, even at levels comparable with current ambient concentrations (Koenig et al., 1979, 1980).

The World Health Organization has synthesized numerous worldwide investigations concerning the health effects of suspended particulate matter and sulphur dioxide in order to suggest the thresholds above which effects on health might be expected among specified populations for short-term and long-term exposures (World Health Organization, 1972, 1979a). Table 2.3 shows that the World Health Organization has determined 250 μg/m^3 as the level for both smoke and sulphur dioxide at which worsening of the condition of patients from short-term exposure might be expected. For long-term exposure, they suggest that 100 μg/m^3 is the level at which adults and children may suffer increased respiratory

Table 2.3 Expected effects of air pollutants on health in selected segments of the population: effects of short-term exposures[a]

Expected effects	24-h mean concentration (μg/m^3)	
	Sulphur dioxide	Smoke
Excess mortality among the elderly or the chronically sick	500	500
Worsening of the condition of patients with existing respiratory disease	250	250

[a]concentrations of sulphur dioxide and smoke as measured by the European acidimetric and reflectance methods respectively.
Source: World Health Organization, 1979a

Figure 2.3 Following the Meuse Valley air-pollution episode in 1930, which resulted in 63 excess deaths, industrial plants have been required to install tall stacks

Source: R. A. Barnes

Table 2.4   Expected effects of air pollutants on health in selected segments of the population: effects of long-term exposures[a]

Expected effects	Annual mean concentration ($\mu g/m^3$)	
	Sulphur dioxide	Smoke
Increased respiratory symptoms among samples of the general population (adults and children) and increased frequencies of respiratory illnesses among children	100	100

[a]concentration of sulphur dioxide and smoke as measured by the European acidimetric and reflectance methods respectively.
Source: World Health Organization, 1979a

symptoms (table 2.4). The World Health Organization reached no firm conclusion as to the effects of TSP because of the limited number of studies available to date, but a tentative figure of 150 $\mu g/m^3$ (annual arithmetic average) for long-term exposure has been suggested.

To protect public health it is considered that a safety factor of two below the figures presented in tables 2.3 and 2.4 would be reasonable. Guidelines for long-term air-quality goals are given in table 2.5. Values for TSP exceed those for smoke because the former includes particles well beyond the respiratory size-range (World Health Organization, 1979a). The inclusion of both respirable fine particulates (0.1 to 2.5 μm) and non-respirable coarse particulates (exceeding 2.5 μm) in TSP levels suggests that it is desirable to issue a separate guideline for fine particulates in the near future.[4]

An additional consideration with regard to fine particulates in health studies is that the type of particle is important. Sulphates, 5 to 20 per cent of suspended particulate matter in urban areas, may be singled out in the future and public-health protection guidelines issued. Sulphates may be responsible for increased asthma attacks, aggravation of heart and lung disease, lowered resistance to respiratory disease in children, and other

---

[4]Fine particles of 0.1 to 2.5 μm are generated as direct combustion products and as gases that are later transformed into particles in the atmosphere. Coarse particles are generated by mechanical events, including wind and friction created by tyres on roads; their sources are largely soil and sand from the Earth's surface. Coarse particles are high in silicates. Fine particles are considered more hazardous to health than coarse particles for the following reasons: (a) deposition of fine particles favours the periphery of the lung (small airways plus air spaces), which for anatomical reasons appear to be especially vulnerable to injury; (b) The clearance of particles from the periphery of the lung is a much slower process than is clearance from the central regions (intermediate and large airways) or upper regions (passages of the head and neck). Slower clearance implies longer exposure, greater dose, and hence, greater risk; (c) Penetration of the superficial tissues and entry into the circulation by particles is likely to be greatest in the periphery of the lung, again for anatomical reasons.

Table 2.5   Guidelines for exposure limits consistent with the protection of public health

	Concentration ($\mu g/m^3$)		
	Sulphur dioxide[a]	Smoke[a]	TSP[b]
24-h mean[c]	100-150	100-150	150-230
Annual arith. mean[d]	40-60	40-60	60-90

[a]Values for sulphur dioxide and smoke as measured by the acidimetric and reflectance methods respectively.
[b]Tentative values suggested for suspended particulate matter as measured by the high-volume sampler widely used in the USA.
[c]Levels indicated for 24-h should not be exceeded on more than seven days a year (equivalent to the 98 percentile for an annual log-normal frequency distribution).
[d]Annual means are specified here in terms of arithmetic means: the corresponding geometric means would generally be a little lower.
Source: World Health Organization, 1979a

air pollution-related conditions. As an indication of the quantifiable effects of sulphates on health, Lave and Seskin (1977) claim that an annual average sulphate concentration of 100 $\mu g/m^3$ increases the mortality rate by 5 per cent.

2.1.2   Effects on Vegetation

Vegetation may be adversely affected by excessive quantities of airborne particles. Particles cover leaves and plug stomata, thereby both reducing the absorption of carbon dioxide from the atmosphere and the intensity of sunlight reaching the interior of the leaf, and suppressing growth of some plants. Specific particles such as fluorides cause additional damage.

Acute injury to plants from sulphur dioxide initially takes the form of bleached patches on broad-leaved plants or bleached necrotic streaking on either side of the mid-vein of parallel veined leaves. Long-term or chronic injury appears as a bleaching of the chlorophyll to give a mild chlorosis or discolouration (yellowing) of the leaf in many plants. In other plants, the bleaching of the chlorophyll reveals the presence of red, brown or black pigments which are normally concealed. Whatever the form of the damage, the result is a reduction in growth and yield. Common plants susceptible to sulphur dioxide pollution include alfalfa, barley, cotton, lettuce, lucerne, rhubarb, spinach and sweet pea. Sulphur dioxide pollution does not always cause damage to vegetation because in sulphate-deficient areas, exposure to low levels of sulphur dioxide may be beneficial to plants by providing the missing sulphur. However, the sulphur dioxide may at the same time reduce the soil pH value, so requiring additional liming. Adverse effects of sulphur dioxide also occur via the effects of acid rain (refer to chapter 4).

One plant particularly susceptible to ambient concentrations of sulphur dioxide is the lichen and it is frequently used as a bioindicator of sulphur dioxide levels. Sulphur dioxide interferes with the photosynthesis in algal cells, eventually destroying the algae's chlorophyll; lichens are especially vulnerable perhaps because they contain relatively little chlorophyll. Experiments involving sulphur dioxide in solution show that different species are affected to different degrees by the same pollutant levels (Hawksworth and Rose, 1976). Sulphur dioxide levels in the atmosphere can be determined by mapping the distribution, percentage cover and size of various species on particular substrates in an area, or from the fate of experimentally-transplanted lichens, or from differences in electrical conductivity between plants, since damaged plants develop weakened cell membranes and lose needed electrolytes. Extensive mapping of lichen species has been undertaken in the United Kingdom (Gilbert, 1974; Hawksworth and Rose, 1970; Morgan-Huws and Haynes, 1973; O'Hare, 1973).

2.1.3   Effects on Materials

Particles soiling fabrics, painted surfaces and buildings add to cleaning and replacement costs by reducing the life of materials and finishes. Particles may cause corrosion either by their intrinsic corrosiveness or by the action of absorbed corrosive chemicals, especially in a moist atmosphere. Soiling of buildings in cities is one of the more obvious manifestations of atmospheric pollution. Particles emitted from diesel exhausts are particularly effective in soiling as they have a high optical absorptance (blackness) and an oily nature (Ball, 1984). Sawyer and Pitz (1983) estimated that the increased soiling that would follow dieselization of 20 per cent of the light duty vehicles in California would lead to an annual welfare loss of $800 million and that this figure related to the cost of domestic households only.

Whereas suspended particulate matter soils or blackens the surface of materials, sulphur dioxide can produce substantial damage to materials. Limestone, sandstone, roofing slate and mortar of buildings and monuments can be severely damaged. The calcium carbonate in limestone and other building materials is readily converted into soluble calcium sulphate (gypsum). The increased volume associated with this chemical change causes scaling, blistering and disintegration of the surface, with the loose material being washed away by rain. The most striking examples of this erosion are the historical monuments of Greece (for example, the Acropolis) and Italy (for example, the Coliseum and the Arch of Titus in Rome). They have withstood the influence of the atmosphere for hundreds or even thousands of years without any great changes, yet in the past few decades alone they have suffered very serious damage.

Table 2.6 Corrosion rates of metals in various environments

Atmosphere	Differential corrosion rate $\mu m\ yr^{-1}$				
	Carbon steel	Zinc	Nickel	Copper	Aluminium
Industrial	40–70	4–6	6	2–5	0.2–1
Urban	20–30	1.7–3.7	1.8	1–2	0.1–0.2
Rural	10	1	0.25	1	0.1

Source: UK House of Commons Environment Committee, 1984

Fabrics, leather, paper, electrical contacts, paint and medieval stained glass are all adversely affected by sulphur dioxide. Human-made textiles such as nylon are especially susceptible to sulphur dioxide or sulphuric acid aerosols. In addition, corrosion of metals, especially iron, steel, zinc, copper and nickel, is accelerated by the presence of sulphur dioxide which encourages the formation of sulphuric acid on the metal surface under moist conditions. Corrosion rates of metals in urban areas may be many times that experienced in rural environments (table 2.6). The lifetime of metals such as galvanized iron and steel can be extremely short in polluted areas with the need for costly replacement or frequent painting of important structures such as electricity pylons and bridges. Thus, for example, a galvanized steel plate with a coating 25 μm thick may require painting after three years in an industrial atmosphere while in a rural atmosphere this may last for 15–20 years before requiring painting.

2.1.4   Atmospheric Effects

One of the most obvious effects of pollutants in the atmosphere is the reduction in visibility caused by the absorption and scattering by solid and liquid aerosols. Relative humidity plays an important role because condensation occurs on hygroscopic particles at humidities as low as 70 per cent: the effect of suspended particulate matter alone (dry haze) is restricted to relative humidities below 70 per cent. During winter, under anticyclonic conditions, the vast quantity of hygroscopic nuclei in an urban area encourages condensation leading to mist and fog formation. Fogs (defined as periods when visibility is less than 1000 metres) form more quickly and persist for longer in a polluted urban atmosphere. However, although fogs are in general more frequent in urban areas, dense fogs – say, with visibilities less than 400 metres – may be less frequent in city centres than in the suburbs. This occurs because fogs with visibility below 400 metres tend to have a higher moisture content than less dense fogs so do not form as readily in the slightly drier and warmer city centre (Chandler, 1976; Musk, 1982).

The presence of vast quantities of suspended particulate matter acting as condensation and freezing nuclei in and downwind of industrial sources and urban areas has been claimed by some to increase precipitation. Changnon (1968), for example, highlighted a significant correlation between annual precipitation at La Porte, Indiana, which lies 48 km downwind of the large heavy-industry complex around Chicago, and the annual number of smoke-haze days in Chicago. The strength of this relationship decreased after the mid-1960s, which may have been due to the decreased emissions of potential condensation and freezing nuclei from industry (Changnon, 1980). However, results from intensive studies such as the METROMEX experiment in St Louis suggest that much of the enhancement of precipitation processes which do take place in and downwind of urban areas is due to dynamic (thermal and mechanical) influences rather than microphysical processes (Changnon, 1970, 1978; Changnon et al., 1977). Studies in and around Greater London also point to pollution nuclei playing an insignificant role in promoting the increased thunderstorm activity which occurs over the metropolitan area (Atkinson, 1968, 1969, 1970, 1971, 1975).

Enhanced cloud and precipitation formation, and especially fog formation, has led to urban areas receiving considerably less than their potential sunshine levels in the past. In Europe and North America, many urban areas during the 1930s and 1940s received only 50 per cent of their potential bright sunshine in winter (Chandler, 1965; Tout, 1979). It was not uncommon for cities such as Pittsburgh and London to require street lighting throughout the day in winter. The losses of solar radiation were greatest at times of low solar elevation – that is, in the early morning and late afternoon hours (Landsberg, 1981). During the past few decades, dramatic reductions in emissions of suspended particulate matter resulting from pollution-control measures, together with social, economic, technological and energy-source changes, have led to substantial improvements in city-centre sunshine levels in major cities (figure 2.4). Although pollution may reduce the amount of bright sunshine or direct solar radiation received in urban areas, this is partially offset by the increase in diffuse radiation which the pollution facilitates (Bridgman, 1981).

Whereas winter visibilities in urban areas have generally improved, summer visibilities have worsened. When relative humidities are below 70 per cent, attenuation of light in the atmosphere by scattering is due to particles of size comparable with the wavelength of the incident light (process of Mie scattering). With visible solar radiation ranging from 0.4 to 0.8 μm, it is the solid and liquid aerosols in the range from 0.1 to 1.0 μm which are responsible for the decrease in visibility. Particles in this size range are principally formed from chemical conversion of gases. Ammonium sulphates and nitrates tend to be prevalent in urban areas since ammonium is a common product of the nitrogen cycle, while sulphates

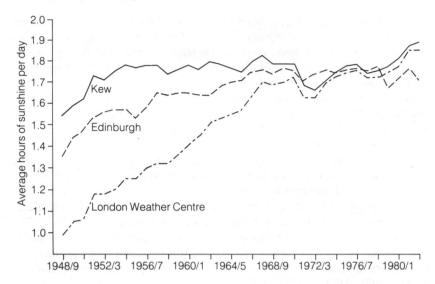

Figure 2.4 Increase in winter sunshine[a] in London and Edinburgh city centres[b]
[a]The figures plotted are 10-year moving averages plotted on the last year, taking winter to be December, January and February.
[b]The hours of winter sunshine are shown for the outer London site of Kew to facilitate comparison with the London Weather Centre in central London.
Source: UK Department of Environment, 1984

and nitrates arise from combustion processes. Sulphates tend to be smaller than nitrates and therefore play the major role in reducing visibility in urban areas. It has been demonstrated in a number of urban areas that as the ammonium sulphate concentration increases, visibility decreases (Eggleton, 1969).

In a study of visibility trends in London and southern England, Lee (1983a, 1983b) has shown that a steady deterioration in summer visibility between 1962 and 1973, especially in the long-range classes, was related to the observed increase in fine particulates especially sulphates (figure 2.5). The increase in sulphate concentrations appeared to be mainly associated with air masses from continental Europe (Barnes and Lee, 1978), although Ball and Bernard (1978a, 1978b) have argued that emissions from Greater London may also be important. After 1973 a sharp improvement in visibility appears to be due to a decrease in sulphate levels related to decreased oil consumption by power stations and industry following the 1973 oil crises.

The transport of sulphates and nitrates over long distances results in regional haze reducing visibility in areas distant from the emission sources of these secondary pollutants. Cass (1979) points to sulphates being the most important single contributor to visibility deterioration in the

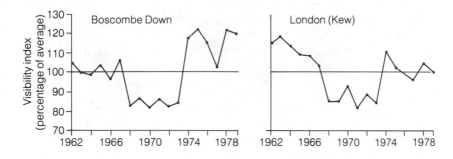

Figure 2.5 Trends in summer visibility at selected sites in southern England 1962-79
Source: Lee, 1983b

northeastern and southeastern USA. The effects of secondary aerosols, especially sulphates, on visibility in the Southwest of the United States were strikingly demonstrated during the nine-month shut down of copper smelters in 1967–8. The copper smelters accounted for more than 90 per cent of the emissions of oxides of sulphur in the Southwest. The effect of the strike on visibility was vividly shown at Tucson, where the number of hazy days was approximately 50 per cent before the strike, but decreased to only 20 per cent during the strike, returned to 50 per cent soon after the strike, and rose to around 80 per cent during the following year when copper production was substantially increased (Trijonis, 1979).

Sulphate and other submicron particles such as hydrocarbons are being increasingly transported further afield to formerly pristine regions of the world such as the Arctic. Thirty years ago there was little evidence of polluted air masses reaching the polar regions, but today, a brownish acid or 'Arctic' haze is commonly experienced in high latitudes, restricting horizontal visibility to between three and eight kilometres (Barrie, 1986; Kemf, 1984). Eurasia appears to be the principal source of the pollution (Barrie et al., 1981; Ottar, 1981; Rahn, 1981; Shaw, 1982). Scientists are concerned not only about the effects on the fragile ecosystem but are also worried about the climatological effects of the pollution. Preliminary research points to the Arctic haze absorbing significant amounts of solar energy during the early spring months. It is possible that this absorption could cause a heating of the lower Arctic atmosphere, causing consequent changes in the global climate.

There is no doubt that global levels of suspended particulates have increased since the turn of the century, but whether this will lead to a global warming or cooling is disputed. Particles over low albedo areas such as oceans may increase the region's albedo and thereby prevent some of the solar energy from being absorbed by the oceans, so causing a surface

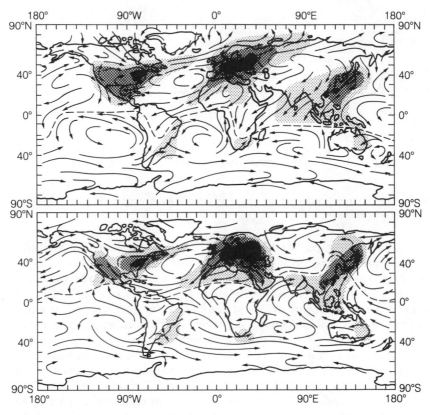

Figure 2.6 Estimated global distributions of anthropogenic aerosols, based on the assumptions that production rate is proportional to the Gross National Product of each country. There is a mean residence time of five days, and surface winds for (a) January, (b) July.
Source: Kellogg, 1978

cooling. However, most of the anthropogenic aerosols occur over land, especially in and around the more industrialized countries, where they are produced such that should they be significantly absorbing, they may reduce the albedo of the region and lead to a surface warming (figure 2.6). With so many uncertainties it is not surprising that little confidence is placed in estimates of the global climatic effect of suspended particulates (Kellogg, 1978).

The most profound effect that suspended particulate matter could have on global climate seems likely to be that which would follow a large-scale nuclear exchange. Research suggests that detonation of thousands of nuclear warheads would cause extensive fires which would pour millions of tonnes of black, sooty smoke into the atmosphere. The fine particulate released would be a very effective absorber of solar energy. Within one to two weeks the particulate would be likely to form a dense smoke pall encircling the mid-latitudes of the Northern Hemisphere. As little as 3 to 5 per cent of solar energy might penetrate this particulate layer for several weeks after the war, resulting in darkness which would only slowly give way to gloomy twilight conditions (Crutzen and Birks, 1982). Such abrupt and sustained reductions in solar energy reaching the surface would probably result in considerably reduced – even subfreezing – temperatures over land masses (Aleksandrov and Stenchikov, 1983; Covey et al., 1984; Elsom, 1984c, 1985; Thompson et al., 1984; Turco et al., 1983). Temperature reductions over oceans would probably be small because of the latters' large heat capacity and rapid mixing of surface waters. The oceans should ensure that the post-war cold period over the continents (the 'nuclear winter' or 'nuclear autumn') would last only a few months. Nevertheless, even a few months of intense cold and twilight conditions would be sufficient to cause widespread damage and destruction of crops and other plants (Ehrlich et al., 1983; SCOPE, 1985). People and animals, even in countries which took no part in the nuclear war, would be likely to die through starvation (the 'nuclear famine') as the effects of the change in climate spread throughout the Northern Hemisphere and possibly into the Southern Hemisphere.

2.2   Photochemical Oxidants

Photochemical oxidants are secondary pollutants produced by the action of sunlight on an atmosphere containing reactive hydrocarbons and oxides of nitrogen (Grennfelt and Schjoldager, 1984). The complex series of photochemical reactions produces various oxidants with the most important being ozone and peroxyacetyl nitrate (PAN) as shown in figure 2.7. Regular measurements of ambient photochemical oxidant concentrations are usually confined to ozone because it is the most plentiful oxidant in a polluted urban atmosphere and the easiest to detect – though probably not the most toxic (Marshall, 1978).

Ozone can form naturally in the atmosphere such that background mean monthly concentrations vary from 0.005 to 0.04 ppm by volume $(10–80 \mu g/m^3)$ depending upon latitude and month of the year. Hourly background values range from 0.005 to 0.05 ppm $(10–100 \mu g/m^3)$. In contrast, ozone levels in urban areas may reach peak hourly concentrations of 0.15 to 0.40 ppm $(300–800 \mu g/m^3)$ as highlighted in table 2.7 and figure 2.8. In some large cities, maximum one hour oxidant concentrations exceed 0.1 ppm $(200 \mu g/m^3)$ on 5–30 per cent of days, while in southern California it is commonplace for peak hourly values to exceed 0.10 ppm $(200 \mu g/m^3)$ on most days of the month between May and September (World Health Organization, 1979b). In 1985, 0.10 ppm of ozone was

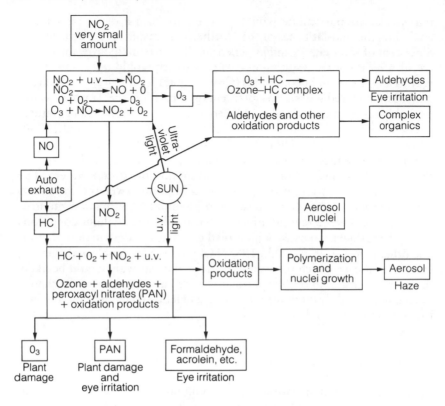

Figure 2.7 Photochemical smog formation
Source: modified from Faith and Atkisson, 1972

Table 2.7   Highest hourly concentration of ozone observed at selected city sites in 1974[b]

City	1-h concentration[a]	
	$\mu g/m^3$	ppm
Riverside, USA	744	0.37
Los Angeles, USA	548	0.27
Eindhoven, The Netherlands	420	0.21
Tokyo, Japan	380	0.19
Osaka, Japan	320	0.16
Washington, D.C., USA	312	0.16
London, UK	294	0.15
Bonn, FRG	290	0.15

[a]0.1 ppm = 200 $\mu g/m^3$.
[b]In Europe, PAN concentrations during summer months are typically 1-2 ppb or less, with isolated peaks during episodes of 16-20 ppb, while in Japan they may reach 30 ppb. In contrast, the Los Angeles basin experiences peaks of 40-210 ppb and monthly means of 4-9 ppb (Temple and Taylor, 1983).
Source: World Health Organization, 1979b

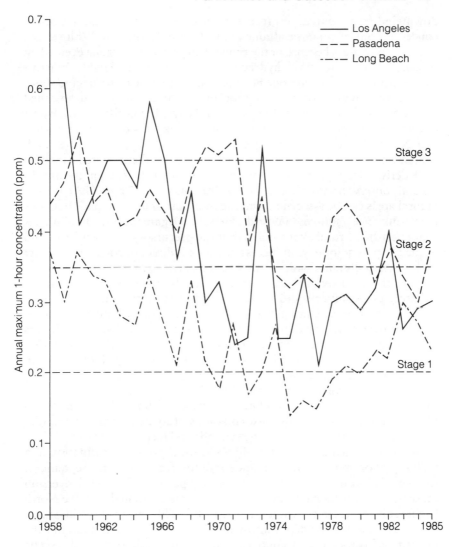

Figure 2.8 Annual maximum hourly ozone concentrations at selected sites in the Los Angeles Basin, 1958–85
Source: Compiled from information supplied by the South Coast Air Quality Management District, El Monte, California

exceeded on 107 days of the year in Los Angeles, on 173 days in Pasadena, and on 166 days in Azusa (US SCAQMD, 1986).

Polluted air masses from urban and industrial areas can affect suburban and rural areas in the direction of the prevailing wind for considerable distances. Indeed, ozone production may only become significant 10 km downwind from sources of precursor emissions, while maximum ozone concentrations may not be attained until 60 km downwind (White et al., 1976). This occurs because of the time taken to produce ozone even when so-called 'highly reactive' hydrocarbons are present. Highly reactive hydrocarbons may require one hour or less for hydrocarbon degradation, less reactive hydrocarbons may take up to three hours, and so-called 'unreactive hydrocarbons' as long as several days. Long-distance transport of ozone and its precursors is becoming increasingly documented in Europe and North America: for example, Cox et al. (1975) found that high concentrations of ozone in southern Eire and the southern United Kingdom were derived from continental European sources 100–700 km distant. In general, observations of hourly concentrations of 0.06 ppm (120 μg/m^3) in rural areas can be associated with the transport of human-made oxidants from distance sources (World Health Organization, 1979b). A characteristic of rural exposures is that the precursors have vanished and ozone can persist for days in succession, since it does not come into contact with other pollutants which act as scavengers.

Photochemical oxidants, as expressed in the form of eye- and nose-irritating pollution episodes or smogs, were first noticed in Los Angeles in the early 1940s. During the 1950s this type of pollution became a problem over much of California but it was not until 1960 that it was recognized as a nationwide problem, and not until the 1970s that photochemical pollution episodes were experienced by large cities in Europe, Australia and Japan.

Southern California, especially Los Angeles, was the first area to experience photochemical pollution because of both the phenomenal increase in motor vehicle exhaust emissions of hydrocarbons and oxides of nitrogen since the 1930s, and the synoptic and meteorological conditions in the state, which are ideal for photochemical pollution formation and build-up. For most of the summer months from May to September, southern California is dominated by a strong persistent low-level inversion caused by compressional heating of subsiding air associated with the North Pacific subtropical anticyclone (Keith, 1980). The height of the inversion determines the mixing depth into which pollutants are emitted, and the intensity of the inversion controls whether any pollutants may escape through the inversion (figure 2.9). When forecasting pollution episodes, the mixing depth is estimated from a tephigram (a graph of the atmospheric temperature profile) by following the dry adiabat through the maximum surface temperature up to the point where it intersects the upper air temperature sounding curve.

Lin (1981) has shown that photochemical pollution episodes in southern California typically begin on days when temperatures aloft at 600–1500 (850 mb.) increase by several degrees, and episodes typically end with a similar temperature decrease. These temperature changes reflect the

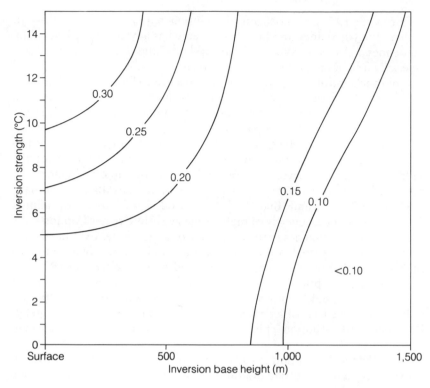

Figure 2.9   Daily ozone concentrations (ppm) as a function of daily inversion strength and height for Azusa, summer 1973
Source: after Keith, 1980

changing intensity of compressional heating by the subsiding air associated with the anticyclone as well as the intensity of the sunlight or solar radiation. Below the inversion the ability of the atmosphere to disperse pollutants horizontally is strongly influenced by the relationship between topography and land- and sea-breezes in southern California. During the day, solar heating of the land, especially inland valleys and south-facing slopes of the mountains which surround the Los Angeles basin, produces a strong temperature contrast with the cooler ocean, thereby giving rise to sea-breezes (Lin and Bland, 1980). In the time required for ozone to be formed from hydrocarbons and oxides of nitrogen, the air is carried inland by the sea-breezes, thereby also transporting the pollution inland too (figure 8.16). When sea-breezes are strong and persistent, higher ozone concentrations may be found inland than occur in the regions which emit the initial pollution. At night-time, a land-breeze reverses the flow of pollutants but ozone concentrations are much lower as photochemical reactions cease after sunset.

During the 1970s, photochemical pollution in major cities of the world, such as Tokyo, Sydney and London, emerged as a problem with startling suddenness. As emissions of hydrocarbons and oxides of nitrogen, especially from motor-vehicle exhausts, increased, so also did the likelihood that photochemical smogs would be experienced. Thus, for example, in the Tokyo Bay area hydrocarbon emissions increased threefold between 1960 and 1970 and oxides of nitrogen increased by a factor of six in the same period. By 1971, Tokyo was experiencing ozone concentrations exceeding 0.15 ppm (300 $\mu g/m^3$) up to ten days per month between April and October (Kagawa and Toyama, 1975). Not all the blame for the sudden emergence of severe photochemical pollution in certain cities can be related to rapid increases in local precursor emissions. Ball and Bernard (1978a, 1978b) found that for the period 1972 to 1976, annual fluctuations in the frequency of high ozone levels in Greater London were largely related to changes in the frequency of anticyclones over northwest Europe which directed ozone from continental Europe over southern Britain. Together with exceptionally hot and sunny summers in 1975 and 1976, these synoptic and meteorological conditions led to London experiencing peak hourly concentrations exceeding 0.20 ppm (400 $\mu g/m^3$), the level of a health-warning first-stage alert in Los Angeles. Photochemical pollution episodes are now a common occurrence in Europe.

Photochemical pollution is today recognized as a worldwide problem for there is no high latitude limitation, as was once believed, to provide the necessary ultra-violet radiation to initiate photochemical reactions. Thus, for example, although the maximum photolysis is less in Oslo (60°N) than Los Angeles (34°N), it is compensated by the long solar days during the summer at the high latitude. Furthermore, high latitude areas such as Scandinavia are frequently receptor areas for oxidants and their precursors transported from distant sources.

2.2.1   Health Effects

The odour (like that of weak chlorine) threshold for ozone is approximately 0.008–0.02 ppm (15–40 $\mu g/m^3$) but the threshold for adverse health effects on the population is about 0.10 ppm (200 $\mu g/m^3$) measured as a one-hour average concentration. Eye irritation occurs at about 0.10–0.15 ppm (200–300 $\mu g/m^3$) with the intensity of eye irritation increasing progressively as concentrations exceed this value (World Health Organization, 1979b). However, other oxidants such as PAN, peroxy-benzol nitrate (PBN) and acrolein are even stronger eye-irritating oxidants than ozone. Hammer et al. (1974) attempted to quantify thresholds for a range of health effects caused by oxidants using a sample of student nurses in Los Angeles. The threshold levels, determined as maximum hourly concentrations, were 0.05 ppm (100 $\mu g/m^3$) for causing headaches, 0.15 ppm (300 $\mu g/m^3$) for eye irritation, 0.27 ppm (530 $\mu g/m^3$) for coughs, and 0.29 ppm (580 $\mu g/m^3$) for chest discomfort. Similar results, but from studies of school children, have been found in Tokyo (Kagawa and Toyama, 1975; Mikami and Kudo, 1973).

The effect of a severe pollution episode on health was clearly demonstrated in the Los Angeles basin in September 1979 when ten consecutive days with peak hourly ozone concentrations exceeding 0.35 ppm (second-stage alert) were recorded (Elsom, 1984a, 1984b). A health survey during this smog revealed 83 per cent of sampled persons reporting discomfort or concern for their health. Fifty-seven per cent of those interviewed complained of burning or irritation of the eyes, while about 25 per cent reported headaches, breathing irritation, sore throats, or stuffy noses. Hospitals reported increases of up to 50 per cent in the number of patients being admitted with chronic lung diseases such as emphysema and asthma. Competitive athletics were cancelled in the affected areas of Los Angeles (McCafferty, 1981).

Epidemiological studies have discovered no increase in mortality with increase in oxidant levels. Only when high temperatures and high oxidant concentrations coincide does daily mortality of the aged increase, and temperature alone is known to cause deaths by hyperthermia. With regard to the effects of ozone on respiratory functions, much of the evidence has been provided by controlled laboratory experiments. Numerous experiments with healthy male subjects exposed to ozone concentrations ranging from 0.1 to 1.0 ppm (200–2000 $\mu g/m^3$) have shown that increased airway resistance and decreased ventilatory performance occurs although there was great variation in individual response. Effects at the lower end of this dose-range were produced when the subjects carried out intermittent light exercise (Bates et al., 1972; Delucia and Adams, 1977; Folinsbee et al., 1975; Hackney et al., 1975; Hazucha et al., 1973; Kerr et al., 1975; Linn et al., 1980; Nielding et al., 1977; Ohmori, 1974).

It is clear that exercise enhances the toxicity effects of ozone by increasing the 'effective dose' received (determined by the product of ozone concentration, mean ventilation volume and duration of exposure). Adams and Schelegle (1983), for example, exposed ten healthy well-trained long-distance runners on six occasions for one hour to ozone concentrations of zero, 0.20 and 0.35 ppm during exercise simulating either training (mean ventilation rate, $V_E = 52$ litres/min) or competition ($V_E = 100$ litres/min). At 0.20 ppm (400 $\mu g/m^3$) the athletes suffered pulmonary function impairment and experienced symptoms including shortness of breath, coughing, excess sputum, throat tickle, raspy throat, and nausea. Exposure to 0.35 ppm (700 $\mu g/m^3$) intensified these effects, resulting in additional pulmonary function impairment and discomfort sufficient to cause premature cessation of exercise for four of the ten subjects and for nine out of the ten to state that they could not have performed at their best in actual competition. Even at 0.20 ppm (400 μg/m^3) exposure, four of the ten subjects stated that, given their post-exposure symptoms, they could not have achieved their best performance. Folinsbee et al. (1978) have shown that whereas at rest ($V_E = 6$ litres/min), no significant effects of ozone were observed in healthy individuals up to and including 0.30 ppm (600 μg/m^3), the 'no effect' threshold for subjects exercising at moderately severe energy expenditure ($V_E = 30$ litres/min) lies somewhere between 0.10 (200 μg/m^3) and 0.30 ppm (600 μg/m^3). At the highest workload ($V_E = 50$ litres/min or more), the 'no effect' level was 0.10 ppm (200 μg/m^3). A confounding factor when the results of different studies are compared is that the subjects may develop a tolerance to oxidants (Hackney et al., 1977a). Thus, for example, Hackney et al. (1977b) showed that controlled laboratory exposures of 0.37 ppm (640 μg/m^3) ozone during intermittent light exercise produced greater clinical and physiological reactivity in Canadians than in southern Californians, who were no more than minimally reactive.

Little has been claimed for long-term or chronic effects on health of elevated ozone concentrations. Animal toxicology studies have shown that chronic exposure to ozone concentrations at and below 0.20 ppm (400 μg/m^3) can produce functional, biochemical and structural changes involving the small airways and adjacent air spaces in animal lungs that are analogous to the changes caused by ageing and early chronic obstructive lung disease (US National Commission on Air Quality, 1981). At higher levels, Last et al. (1979) have shown that fibrosis, or scarring of the lungs, may occur when subjects are continuously exposed to ozone concentrations of 0.50 ppm (1000 μg/m^3) or more, and that the predicted threshold for these changes is in the range of current Los Angeles ambient levels. Similar effects on people cannot be either inferred or rejected on the basis of short-term controlled human studies and they appear to lie beyond the limits of detection by conventional epidemiological techniques.

From community and controlled human exposure studies, the World Health Organization (1979b) concluded that the first adverse effects in people appear at ozone concentrations of 0.10 to 0.25 ppm (200–500 μg/m^3). As a result, the long-term goal for the protection of public health was recommended to be a maximum one-hour oxidant concentration of 0.06 ppm (120 μg/m^3) or a maximum eight-hour concentration of 0.03 (60 μg/m^3). This provides little or no safety factor in view of the relatively high natural concentrations of ozone and the long-distance transport of ozone. Nevertheless, the World Health Organization urged that every effort should be made to adopt pollution control strategies for achieving the proposed guideline or, at least, for not exceeding it more than once a month.

2.2.2   Effects on Materials and Vegetation

Ozone, an oxidant much stronger than oxygen, causes cracking of stretched rubber at hourly concentrations of only 0.01–0.02 (20–40 μg/m^3), although ozone inhibitors can be built into rubber products such as vehicle tyres and rubber insulation. Ozone also attacks the cellulose in textiles, reducing the strength of such items and all oxidants cause some fading of fabrics and dyes. Textile fabrics affected include cotton, acetate, nylon and polyester. Erosion of exterior painted surfaces due to attack on the organic binder in the paint also occurs. Even inside buildings such as museums and art galleries, where ozone levels may be half of those found outside, there is concern that the organic materials in antiquities might be damaged (Shaver et al., 1983; Thompson, 1978).

Oxidants cause acute and chronic injury to plants, causing necrotic patterns on leaves, growth alterations, reduced yields and reductions in the quality of the plant products. Typical effects of ozone are stippling or flecking (brown spots or flecks which subsequently turn white) on the upper surface of leaves. Ponderosa pine trees seem to be especially sensitive (Miller et al., 1969) and in the San Angeles and San Bernadino forests of southern California, vast areas of trees are dying as the result of prolonged exposure to photochemical oxidants. The trees are injured by the high ozone concentration and become very susceptible to fatal pest outbreaks of pine-bark beetles. On the western slopes of the southern Sierra Nevada Mountains, where ozone levels are much lower, widespread injury by chlorotic mottle on the needles and more advanced stages of chlorotic decline have been attributed to the smog. Similar damage has been observed in other regions of the United States, in Mexico and in Israel, while in many parts of Europe die-back of forests appears to be happening on a massive scale. Air pollutants, including ozone, are considered to be largely responsible for the damage to European forests (Hinrichsen, 1986; Skarby and Sellden, 1984).

It is estimated that direct commercial losses from ozone damage in the United States add up to several billion dollars a year (Heck et al., 1982, 1983). This estimate is based on crop response data from the National Crop Loss Assessment Network which began in 1982 and covers all of the major crops. For California alone, total crop losses are estimated to reach about one billion dollars a year. Other crops especially sensitive to ozone damage include spinach, tomatoes, pinto beans and tobacco. The tobacco (*Nicotiana tabacum*) strain Bel-W$_3$ may be employed as a bioindicator of ozone levels in excess of 0.05 ppm (100 μg/m^3) as this is the level at which necrotic damage to the leaves occur in this species (figure 2.10; Bell and Cox, 1975). In general, an ozone concentration of 0.05 ppm (100 μg/m^3) for a four-hour period or 0.03 ppm (60 μg/m^3) for an eight-hour period appears to

Figure 2.10   Typical symptoms of ozone damage on the tobacco plant, Bel-W$_3$
Source: M. R. Ashmore

be the threshold of damage for sensitive plants. However, if even low concentrations of sulphur dioxide are present, the time for the effects to occur is reduced. In addition, 0.01 ppm (20 μg/m^3) of PAN for a six-hour exposure or 0.015 ppm (30 μg/m^3) of PAN for a four-hour exposure has been observed to cause undersurface glazing, silvering or bronzing of new leaves of sensitive plants (Temple and Taylor, 1983).

2.2.3   Atmospheric Effects

Secondary pollutants formed during the chemical reactions that create photochemical pollution episodes cause a marked reduction in visibility–hence the use of the term 'smog'. The smog varies in colour according to the nature of the various constituents of the smog, being brownish in Los Angeles, whereas the London smogs of 1976 were of a greenish hue. Ozone also plays a part in producing regional hazes by promoting the oxidation of sulphur dioxide to sulphate particles. Ozone itself is virtually a colourless gas (very faintly blue).

Increased anthropogenic emissions of the precursors of ozone formation have led to increasing tropospheric ozone concentrations in the middle latitudes of the Northern Hemisphere (Angell and Korshover, 1980). Fishman et al. (1979) have calculated that infra-red radiation absorption by ozone has led to the average temperature of the Northern Hemisphere increasing by 0.2°C. If tropospheric ozone concentrations were to double, it is estimated that global surface temperature would increase by 0.7° to 0.9°C (Hov, 1984; Lal et al., 1986).

2.3 Oxides of Nitrogen

Oxides of nitrogen (NO_x) are produced by natural processes, including bacterial action in the soil, lightning and volcanic eruptions and by human activity during combustion processes at temperatures higher than about 1000°C. Nitric oxide (NO) and nitrogen dioxide (NO_2) are the most important oxides of nitrogen for pollution studies because other oxides of nitrogen such as nitrous oxide (N_2O), denitrogen trioxide (N_2O_3), denitrogen tetroxide (N_2O_4), denitrogen pentoxide (N_2O_5), and nitric acid (HNO_3) vapour, which may exist in ambient air, are not known to have any biological significance (Lee, 1980; World Health Organization, 1977a). The principal emissions of oxides of nitrogen from human activities are from the combustion of fossil fuels in stationary sources (heating, power generation) and in motor vehicles (internal combustion engine). The relative contribution of each source varies from country to country in relation to differences in fuel use (Derwent and Stewart, 1973; World Health Organization, 1977a).

2.3.1 Health Effects

Much of the attention paid to oxides of nitrogen in urban areas relates to their significant role in promoting photochemical pollution. Nevertheless, oxides of nitrogen may cause adverse health effects on their own, as well as when other pollutants are present. Annual mean concentrations of nitrogen dioxide, monitored as the indicator of levels of gaseous oxides of nitrogen, in urban areas throughout the world are typically in the range of 20–90 $\mu g/m^3$ (0.01–0.05 ppm) with maximum 24-h means being typically between two and five times greater than annual means and maximum hourly means between five and ten times greater than annual means. However, in some areas, such as near industrial plants producing nitric acid or explosives, or near power stations, very high nitrogen dioxide levels may occur. Exceptionally high nitrogen dioxide concentrations may also occur indoors from sources such as gas-fired heaters, boilers and cookers as well as from cigarette smoking (refer to Chapter 1, section 1.9). An example of an area suffering high levels of nitrogen dioxide due to emissions from a TNT plant is that of Chattanooga, Tennessee. From 1968 to 1973 significantly higher respiratory illness rates in resident families were revealed compared with families in less-polluted areas of the community (Shy and Love, 1980).

As the mean annual nitrogen dioxide concentration decreased from 190 μg/m^3 in 1968 to 94 μg/m^3 in 1972 and to 60 μg/m^3 in 1973, the excess of respiratory illness in the high-pollution area reduced from 18.8 per cent in 1968 to 10.3 per cent in 1973. The lower nitrogen dioxide concentration in 1973 reflected the effect of a two-month labour strike at the munitions plant and the excess of illness during that period was reduced by about one-third compared with 1972. Subsequently it was concluded that the 90th percentile 24-h concentration of 600–900 μg/m^3 of nitrogen dioxide is associated with increased illness.

Apart from enhanced susceptibility to respiratory infections, nitrogen dioxide may lead to increased airway resistance and increased sensitivity to bronchoconstrictors in sensitive individuals. Kerr et al. (1978) showed that seven in thirteen asthmatics, one in seven bronchitics, and one in ten normal subjects reported chest tightness, burning of the eyes, headache or dyspnea with exercise at 940 μg/m^3 (0.5 ppm): the symptoms were said to be mild and were unaccompanied by objective evidence of decreased lung function. Orehek et al. (1976) found that at a 1-h nitrogen dioxide exposure level of 210 μg/m^3 (0.11 ppm) 13 of 20 slight to mild asthmatics experienced substantially more bronchoconstriction in a carbochol provocation test than without the nitrogen dioxide. This finding suggests that asthmatic subjects or other similarly sensitive subgroups might be at risk for adverse effects on pulmonary performance at nitrogen dioxide levels well below those previously found to be innocuous in healthy subjects (Nielding and Wagner, 1979; Orehek et al., 1976). However, a similar study by Hazucha et al. (1983) failed to substantiate these findings.

From the limited number of epidemiological studies, together with occupational exposure studies and laboratory experiments with animals and humans, the World Health Organization (1977a) recommends that the maximum exposure level of nitrogen dioxide for the protection of public health should be 190–320 μg/m^3 (0.10–0.17 ppm) for one hour, not to be exceeded more than once per month. This standard incorporates a minimum safety factor of between three and five, as a nitrogen dioxide concentration of 940 μg/m^3 (0.5 ppm) has been selected as an estimate of the lowest level at which adverse health effects resulting from short-term exposure can be expected to occur.

2.3.2   Effects on Vegetation

Oxides of nitrogen rank second to sulphur compounds in their contribution to acid rain which may affect terrestrial and aquatic ecosystems. In northeastern United States, for example, 30 per cent of the acidity of precipitation (below pH 5.6) is believed to be caused by the oxides of nitrogen producing nitric acid, 65 per cent to sulphuric acid and 5 per cent to hydrochloric acid. However, whereas the contribution of sulphate to the problem of acid precipitation is levelling off, that of nitrate is increasing (Babich et al., 1980). The subject of acid rain is explored more fully in chapter 4. In addition to the indirect effects of oxides of nitrogen via acid precipitation on vegetation, prolonged exposure to nitrogen dioxide concentrations of 470–1880 $\mu g/m^3$ (0.25–1.0 ppm) may suppress growth of such plants as tomatoes, pinto beans and navel oranges (Wark and Warner, 1981).

2.3.3 Atmospheric Effects

Significant atmospheric effects of oxides of nitrogen include their role in reducing visibility and their potential for causing a global surface temperature increase. In the atmosphere, nitric oxide emissions are readily converted to nitrogen dioxide and nitrate aerosols, both of which reduce visibility. Nitrogen dioxide absorbs visible light (and strongly absorbs ultraviolet radiation), and at a concentration of 470 $\mu g/m^3$ (0.25 ppm) will cause an appreciable reduction in visibility. Generally, nitrogen dioxide appears to be less important than nitrates in causing visibility reductions, while nitrates currently appear to be less important than sulphates because of lower emissions and because of the variable size range of nitrates. Thus, for example, although nitrates were measured to be mostly in the all-important fine particulate fraction (less than 2.5 μm) in Denver and Los Angeles, studies in Detroit and New York showed most of the nitrates to lie within the coarse particulate fraction (2.5 μm and above). Even so, long-distance transport of fine nitrates contributes to the regional hazes increasingly experienced since the 1960s in areas such as southwestern and northeastern United States as well as in Europe.

If the global atmospheric concentration of nitrous oxide increases, this may lead to a surface warming because of the 'greenhouse effect' – that is, the ability of nitrous oxide to absorb infra-red radiation. Nitrous oxide is mostly maintained at its present concentration by biological decay and conversion processes taking place in soil and in the oceans–processes referred to as 'de-nitrification' (Kellogg, 1978). It has been suggested that the increasing use of nitrate fertilizers by humankind may accelerate the biological production of nitrous oxide and raise its atmospheric concentration (Crutzen, 1976; McElroy et al., 1976). The amount of this increase in nitrous oxide levels is still uncertain, since estimates vary from a trivial increase to as much as a factor of two in the early part of the next century. The latter increase is likely to produce a global surface temperature rise of between 0.5°C (Yung et al., 1976) and 0.7°C (Hov, 1984). An increase of nitrous oxide concentration in the atmosphere may also lead to a decrease in the stratospheric ozone concentration which may also affect surface temperatures. The potential effect of oxides of nitrogen on stratospheric ozone is explored more fully in chapter 6.

2.4   Carbon Monoxide

Carbon monoxide is a colourless, odourless gas produced by the incomplete combustion of carbon-containing fuels and by some biological and industrial processes. Every year human activities may put some 1500 teragrams (1 $Tg = 10^{12}g$) of carbon monoxide into the Earth's atmosphere compared with the 1200 Tg/year from natural sources (US National Research Council, 1977a).

The major source of carbon monoxide emissions at breathing level outdoors is the exhaust of petrol-powered motor vehicles; the diesel engine (compression ignition), when properly adjusted, emits little carbon monoxide. Locally, high concentrations of carbon monoxide may occur near industrial plants such as power stations, petroleum refineries, iron foundries and steel mills, as well as in the vicinity of refuse burning, whether in incinerators or openly. Carbon monoxide concentrations in urban areas are closely related both to motor-traffic density and weather conditions. Carbon monoxide levels show a distinct diurnal pattern with peaks corresponding to the morning and evening traffic rush-hours. Levels of this pollutant decrease very rapidly with distance from emission sources (figure 2.11). Data from the United States and Japan show that 8-h mean concentrations of carbon monoxide are generally less than 20 mg/m^3 (17 ppm) although maximum 8-h mean concentrations of up to 60 mg/m^3 (53 ppm) have occasionally been recorded. Short-term concentrations of even higher concentrations have been observed in confined spaces such as tunnels, garages, loading bays, underpasses, underground car parks, and in narrow congested roadways (Bodkin, 1974; World Health Organization, 1979c; Wright et al., 1975). It should be noted, however, that at normally encountered levels, there are no known adverse effects of carbon monoxide on vegetation and materials.

2.4.1   Health Effects

Carbon monoxide is absorbed through the lungs and reacts with haemoproteins, especially with haemoglobin of the blood. This in turn results in a reduction of the oxygen carrying-capacity of the blood, and also interferes with the release of the oxygen which is carried to the tissues. Carbon monoxide has an affinity for haemoglobin that is 200–240 times greater than that of oxygen, and carboxyhaemoglobin (COHb) is therefore a more stable compound than oxyhaemoglobin. Carbon monoxide is not a cumulative poison, but is excreted or absorbed, depending upon the level of carbon monoxide in the ambient air, the amount of carboxyhaemo-globin in the blood, barometric pressure, the duration of the exposure, and the rate of ventilation of the lungs (which may be related to exercise).

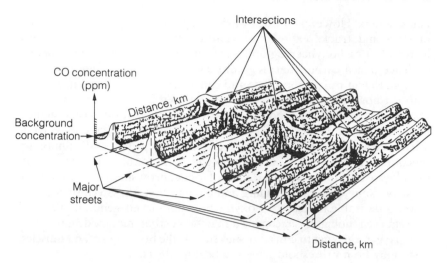

Figure 2.11   Inferred spatial distribution of carbon monoxide concentrations in an urban area
Source: Ott, 1977

The World Health Organization (1972) has calculated these inter-relationships assuming the individual has a basal carboxyhaemoglobin content and is breathing ambient air at sea-level and that 50 per cent of the equilibrium is reached after about three hours (table 2.8). A useful approximate relationship relating ambient carbon monoxide concentration to carbon monoxide saturation of the blood is that for every 1 ppm of carbon monoxide with which the body is in equilibrium (by diffusion through the alveolar sacs), 0.165 per cent of the body's haemoglobin will be combined in the form of carboxyhaemoglobin.

By far the most common cause of high carboxyhaemoglobin concentration in people is the smoking of tobacco and the inhalation of the products by the smoker. Typically, cigarette smokers generally have a mean carboxyhaemoglobin level of 5 per cent compared with 1 per cent in

Table 2.8   Interrelationships between the level of carbon monoxide in the ambient air, the amount of COHb in the blood and the duration of exposure

Ambient CO		Carboxyhaemoglobin %		
ppm	*mg/m³*	*after 1 hour*	*after 8 hours*	*at equilibrium*
100	117	3.6	12.9	15
60	70	2.5	8.7	10
30	35	1.3	4.5	5
20	23	0.8	2.8	3.3
10	12	0.4	1.4	1.7

Source: World Health Organization, 1972

non-smokers. However, traffic policemen, garage attendants, and drivers of taxis and trucks exposed to vehicle exhaust emissions experience increases of carboxyhaemoglobin levels up to about 3 per cent. There is evidence that a carboxyhaemoglobin level of between 1 and 2 per cent affects behavioural performance and can aggravate symptoms in patients with cardiovascular disease; a level of between 2 and 5 per cent causes impairment of vigilance and time-interval discrimination, visual acuity, brightness discrimination, and certain other psychomotor functions; while a level exceeding 5 per cent is associated with cardiac and pulmonary functional changes. As levels increase above approximately 10 per cent, the adverse effects include headaches, fatigue, drowsiness, reduced work capacity, coma, respiratory failure and ultimately death. Given such effects it seems desirable to keep carboxyhaemoglobin levels generally below 2 per cent even though it is recognized that with carbon monoxide interfering with oxygen-carrying to crucial tissues such as the brain, heart and muscles, there may be no threshold level for health effects.

The World Health Organization (1972) has suggested a long-term goal of $10 \, mg/m^3$ (9 ppm) for an 8-h period and $40 \, mg/m^3$ (35 ppm) for a 1-h averaging period. The 8-h averaging is employed because it takes from four to twelve hours for the carboxyhaemoglobin level in the human body to reach equilibrium with the ambient carbon monoxide concentration. One limitation in using an ambient air quality standard to protect public health, even among non-smokers, is that personal activities such as cooking with gas stoves, or travelling in or working with motor vehicles, produce a carbon monoxide exposure which is only poorly reflected by outdoor monitoring stations (Wallace and Ziegenfus, 1985).

2.4.2   Atmospheric Effects

Khalil and Rasmussen (1984) provide measurements for the past few years at Oregon which indicate that the background concentration of carbon monoxide is increasing at a rate of on average approximately 6 per cent per year. Human activities are the likely cause of a substantial portion of the observed increase. However, because of the short atmospheric lifetime of carbon monoxide and the relatively few years of observations, fluctuations of sources and sinks related to the natural variability of climate may have affected the observed trend. Increased carbon monoxide may deplete tropospheric hydroxyl (OH) radicals, thereby reducing the yearly removal of dozens of natural and anthropogenic trace gases. Consequently, Khalil and Rasmussen (1984) suggest that increased carbon monoxide may indirectly intensify a global warming. Perturbation of the stratospheric ozone layer may also result as the lifetimes of ozone-scavenging trace gases may be increased.

2.5 Toxic Metals: Lead

Five metals found in the air – namely, beryllium, cadmium, lead, mercury and nickel – represent potential or real public-health hazards. With the exception of lead, these toxic metals only generally give rise to concern for the public living near to specific toxic metal sources (industrial plants, waste dumps) and for the labour force of certain metal-refining or production industries. What makes lead exceptional compared with other toxic metals is that humankind uses vast quantities of this metal and it is widely dispersed throughout the environment (UK Royal Commission on Environmental Pollution, 1983). The major sources of lead in the environment which are of significance for the health of communities arise from the industrial and other technological uses of lead. The major dispersive non-recoverable use of lead is in the manufacture of alkyllead fuel additives. Other important sources of environmental lead relate to the manufacture of batteries, sheet and pipe, cable sheathing, solder, shot and paint.

Alkyllead compounds (tetraethyllead and teramethyllead) have been used as anti-knock additives in petrol (gasoline) for over 50 years. Their use increased steadily into the 1970s, after which the introduction of regulations on the maximum permissible concentration of lead in petrol (especially in the USA), together with the world oil and energy crisis, led to a decline in the amount of lead used. Lead compounds are added to fuel to prevent knocking, the distinctive 'pinking' noise which occurs when there is spontaneous ignition of the petrol and air mixture in the corner of the cylinder furthest away from the spark plug (the end-gas zone). For a given design of engine, the presence or absence of knocking is determined by the combustion quality or 'octane number' (percentage of a very high knock-resistant type of hydrocarbon) of the petrol used: the higher the octane number, the higher the compression ratio at which the engine can be run without knocking. The addition of organic lead compounds (lead alkyls) is a cheaper and more convenient method of boosting the octane quality of petrol than by intensive refining. The presence of lead in the fuel has the effect of delaying the abnormal state of oxidation in which the end-gas ignites, and hence allows engines to operate at higher compression ratios and under greater loads before knocking occurs. If tetraethyllead and teramethyllead are not employed, the choice to the motor industry is either to accept increasing refining costs of the petroleum (to increase the proportion of knock-resistant hydrocarbons) or to make use of reduced octane quality in vehicles (engines such as the diesel and stratified charge do not knock, and do not therefore need high octane fuel).

Two factors lie behind the actions which various countries have taken to reduce or eliminate lead in petrol–the concern about the direct effects on health of airborne lead, and the control of other gaseous emissions through the use of catalytic devices which do not function properly with leaded petrol. Catalytic devices are currently employed in the USA, Japan, and Canada, and have been proposed in several other countries. Following combustion of motor fuel containing alkyllead additives, over 70 per cent of the lead immediately enters the atmosphere, the rest being trapped in the crank case oil and in the exhaust system of vehicles. From studies in the Log Angeles basin, Huntzicker et al. (1975) estimated that of the lead emitted into the atmosphere, 40 per cent becomes near fallout (on or near roads), 8 per cent is deposited within the metropolitan area, and 24 per cent is more widely dispersed, having a residence time of one or two weeks. Evidence for widespread dispersal of lead comes from studies in Greenland where lead deposits in ice layers of the late 1960s reached concentrations 400 times greater than the natural background (Murozumi et al., 1969).

The concentration of lead in air may vary from 2 to 4 $\mu g/m^3$ in large cities with dense motor traffic, to less than 0.2 $\mu g/m^3$ in most suburban areas and still less in rural areas. Concentrations are highest along highways during rush hours where daily levels of between 14 and 25 $\mu g/m^3$ may be reached (World Health Organization, 1977b). Excessive concentrations may also be measured in the vicinity of industrial sources such as lead smelters: for example, at El Paso, Texas, a measurement of 80 $\mu g/m^3$ was taken in the immediate vicinity of a large ore smelter, although concentrations fell rapidly with distance, reaching 1 $\mu g/m^3$ about 5 km away (World Health Organization, 1977b). Many nations have adopted an annual mean lead concentration of 2 $\mu g/m^3$ as a long-term health-related goal for air-pollution control purposes.

People are exposed to varying levels of lead in the atmosphere according to where they live and work and how they commute. However, the contribution of airborne lead by direct inhalation is but one pathway by which airborne lead contributes to the total lead intake in people (figure 2.12). In addition to direct or indirect airborne lead contributions, there are several other ways in which lead levels in the body are increased. In areas where the water is soft (low in calcium and magnesium) and where, at the same time, lead pipes and lead-lined water storage tanks are used, a percentage rise in the body burden of lead occurs, which is reflected in elevated values of lead in the blood. Since lead in water ingested independently of food is more readily absorbed, it may provide a relatively greater contribution to the blood lead level than lead in food. Lead concentration in food is highly variable, with foods and wines which are stored in lead-soldered cans or lead-glazed pottery being especially high in lead content. For infants and young children, an additional source of lead may come from miscellaneous lead-containing objects which are licked, chewed or eaten, as well as from hand-to-mouth contact with lead-contaminated dust. Lead contamination of dust may be derived from

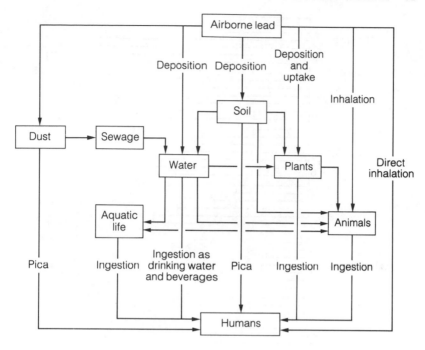

Figure 2.12 Contribution of airborne lead to total lead intake
Source: World Health Organization, 1977b

deposition of lead from vehicle exhausts (Millar and Cooney, 1982), from lead-based paint in old houses, or in the case of some locations, as a result of fallout from lead smelters. Inner-city children appear to be particularly prone to experience high blood lead concentrations as a result of the ingestion of highly lead-contaminated dust in and around traffic congested streets (Wilson, 1983).

Levels of blood lead in urban populations vary markedly between countries. Average blood lead concentrations range from 6 μg/dl in Peking and Tokyo, through 10 to 20 μg/dl in western European and Indian cities, to as much as 22 μg/dl in Mexico City (UK Royal Commission on Environmental Pollution, 1983). Substantial reductions in some countries have been achieved as, for example, in the United States, where mean blood lead concentration decreased from 15.8 μg/dl in 1976 to 10.0 μg/dl in 1980 in line with reductions in petrol lead.

Clinical signs of lead poisoning are fairly well established. Anaemia is a characteristic early toxic effect in people, with a slight decrease in the haemoglobin levels occurring at a mean level of about 50 μg/dl in adults but at approximately 40 μg/dl for some population groups and particularly in iron-deficient children (Betts et al., 1973; Peuschel et al., 1972).

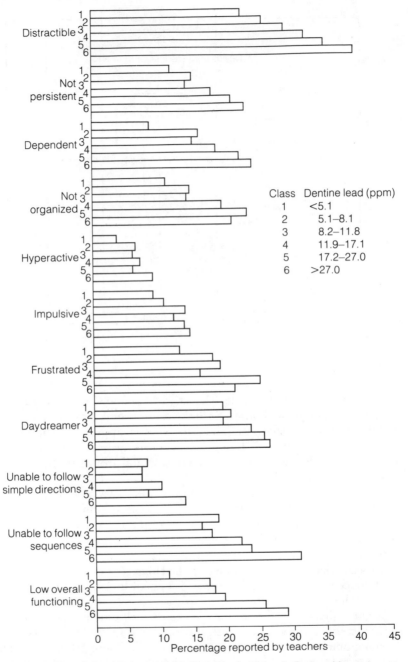

Figure 2.13   The relationship between lead in teeth of children in Boston, Massachusetts, and their behavioural and educational difficulties
Source: Needleman et al., 1979

Dysfunction of the brain occurs with levels of 50–60 μg/dl in children and 60–70 μg/dl in adults, while acute or chronic encephalopathy and damage to the kidney occur at levels of 60–70 μg/dl in children and approximately 80 μg/dl in adults (World Health Organization, 1977b).

While abnormally high blood lead levels produce unmistakable lead poisoning symptoms, the effects of lower blood lead levels, especially in children, are less obvious and are the subject of controversy. The current concern for young children is that blood lead levels even as low as 10 μg/dl may be causing subtle neurological damage in some susceptible children without ever exhibiting classical signs of lead encephalopathy. These altered behaviours may be recognized by parents, teachers, and clinicians as attentional disorders, learning disabilities, or emotional disturbances which impair progress in school (Needleman and Shapiro, 1974). Although Burde and Choate (1972) and Peuschel et al. (1972) observed dysfunction of the central nervous system (irritability, clumsiness, fine motor dysfunction, impaired concept formation, etc.) in groups of children, whose blood lead levels were in all cases above 40 μg/dl, an increasing number of subsequent studies have produced evidence of altered neuropsychological behaviours in children in relation to much lower lead levels. Needleman et al. (1979) studied two working-class towns near Boston, Massachusetts and compared the tooth lead levels (teeth provide a long-term record of lead exposure whereas blood lead levels indicate current and recent exposure) of 2146 children with their teacher's rating on eleven behavioural and educational characteristics. A clear association was evident between high lead levels and negative ratings (figure 2.13). Lansdown et al. (1983) confirmed Needleman's findings in England in a group of Greenwich school children. In addition, the children with the higher lead levels had a statistically significant average IQ deficit of seven points. Further support for these findings came from research in Philadelphia by Marecek et al. (1983), in New York by David et al. (1983) and in Germany by Winneke (1983), yet some researchers are reluctant to accept these findings.

The increasing evidence for the harmful effects of lead, particularly in children, at levels once thought to be 'safe' has resulted in several countries seeking ways of reducing lead in the environment so as to increase the safety margin for protecting public health (UK Royal Commission on Environmental Pollution, 1983). One of the obvious sources of lead amenable to control is that of lead in petrol because lead consumption had been steadily increasing during the past few decades and because vehicles emit lead in close proximity to people. Whether this pollution-control approach is effective in significantly reducing lead levels in people depends upon the contribution of petrol lead to lead uptake in the body.

The relative proportions of sources which contribute to the total lead intake by adults and children are both variable and disputed. The claim by the Lawther Report (UK Department of Health & Social Security, 1980) that only 10 per cent of the total lead intake in children ultimately comes from vehicle emissions of lead has received increasing scepticism. Two studies have suggested that this figure is a gross underestimate. An Italian study, the Isotopic Lead Experiment, undertaken in and around Turin, northern Italy, was sponsored by the EEC (Commission of the European Communities, Joint Research Centre, 1982). In the area around Turin, the lead added to all petrol distribution between 1975 and 1980 had an isotopic ratio distinct from that used beforehand and afterwards, being derived from the Broken Hill mine in Australia. This experiment enabled estimates of the pathways taken by petrol lead in the environment and of the total contribution of petrol lead to the body burden. The percentage of isotopic lead intake averaged 24 per cent in Turin adults, compared with 11 to 12 per cent in adults living 25 km or more from the Turin centre. Estimates of the special lead intake in children reached 30 per cent after 1979. To assess the true proportion of petrol lead to the total lead intake, one should add the petrol lead that had already accumulated in the body prior to the experiment and also the lead from petrol obtained from outside the sample area.

Even higher contributions to blood lead levels of the lead originating from vehicle exhausts is suggested by Annest et al. (1982) who analysed samples of blood from 9933 Americans. Between 1976 and 1980, 46 per cent of the blood lead level of the average American was shown to originate from petrol lead emissions. Black children had higher blood lead levels than older blacks or white as a result of exposure to the same levels of petrol lead emissions, and an estimated 62 per cent of their blood lead was caused by petrol lead emissions. During the study period, the levels of lead in people's blood decreased by 37 per cent whilst lead consumed in petrol decreased by 56 per cent (figure 2.14). The closeness of this association is increased when it is realized that dietary lead intake during this period increased by 14 per cent. Clearly, the improvement in blood lead levels was a result of the reduction in the lead content of petrol and to the increased use of unleaded petrol (US Environmental Protection Agency, 1983b). Rabinowitz and Needleman (1983) provide additional support for interpreting the primary cause of the fall in blood lead concentration as being due to the reduction in petrol lead. They found a significant correlation between the lead content of all petrol sold in Massachusetts between April 1979 and April 1981 and the concentration of lead in umbilical cord blood from births at a Boston hospital.

In conclusion, it increasingly appears that petrol in lead contributes a significant and sizeable proportion of the lead intake of adults and children. Not surprisingly, strong public and medical pressure is being exerted on governments in many countries to reduce the amount of lead added to petrol and to increase the proportion of motor vehicles able to run on unleaded petrol.

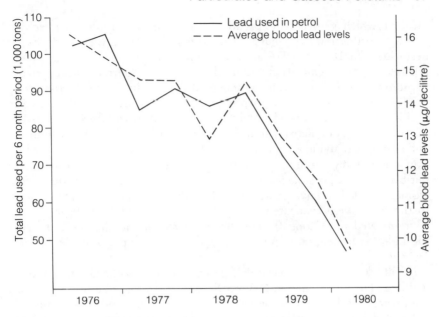

Figure 2.14 Lead used in petrol production and blood lead levels in the USA, February 1976–February 1980
Source: Wilson, 1983

2.6 Toxic Chemicals

There are between 60 000 and 70 000 synthetic chemicals available today, and between 200 and 1000 new chemicals are introduced each year (Shaikh and Nichols, 1984). Although the acute effects of many of these chemicals have been recognized, little or no knowledge of their potential chronic (long-term) effects on people (for example, cancer initiated by a chemical may take 15 to 40 years to develop) or the environment is available. An enormous amount of work needs to be undertaken to identify those chemicals which may cause chronic adverse effects before these pollutants become widely distributed in the environment.

Toxic chemicals are released into the atmosphere not only from major industrial point sources emitting large quanitities of one or two chemicals, but also from small commercial and residential facilities, mobile sources, industrial incinerators, waste disposal facilities and sewage treatment plants emitting numerous toxic chemicals, often of unknown identity. Public recognition of the potential adverse health effects of toxic chemicals has been prompted by a number of incidents associated with chemical plants which manufacture solvents, pesticides, herbicides and plastics. One incident which gained widespread international attention occurred at Seveso, northern Italy, where trichlorophenol (TCP) was made as a herbicide. On 10 July 1976 an explosion at the Seveso chemical plant released various chemicals including dioxin (2,3,7–8 tetrachlorodibenzo-paradioxine or TCDD), which is accepted as probably being carcinogenic to humans, into the atmosphere. The toxic cloud of dioxin, trichlorophenol, ethylene glycol and caustic soda spread downwind over the surrounding populated region (figure 2.15). Within two weeks, animals and plants were dying and people were being admitted into hospital with skin lesions and vomiting. No deaths were linked directly with the disaster but because of a possible link between dioxin and genetic mutations, 90 women decided to have abortions. More than 700 people living closest to the plant were evacuated (zone A). Another 5000 people in a less contaminated area (Zone B) were allowed to stay in their homes, but were forbidden to garden, raise animals or let their children play outside (Fuller, 1977).

It had taken nearly two weeks before evacuation and effective safeguards had been introduced in and around Seveso because the authorities were not aware that dioxin was involved. The information had been available to the chemical company but it had not been communicated to the relevant authorities. As a result of the Seveso experience the European Community introduced the 'Seveso directive' in 1984. It obliges companies that manufacture or use a wide range of dangerous chemicals to identify the risks present in their factories and to tell workers and local residents what

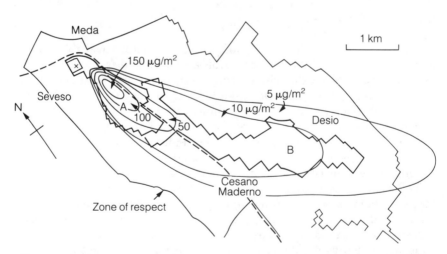

Figure 2.15 Calculated deposition of dioxin in the Seveso area assuming the release of 2 kg of dioxin. Zone A was evacuated on 24 July 1976, while zone B was subsequently decontaminated
Source: Warner, 1979

those risks are. The directive also lists the amounts of dangerous chemicals, such as chlorine, ammonia, and so on, that can be stored safely within 500 metres of each other. It provides 'threshold' quantities of other dangerous chemicals: if a factory stores more than the 'threshold' amount, its owners must give the authorities precise details of the plant and the safety measures they take.

Similar 'community right-to-know' laws, together with requirements for emergency plans and response programmes, are being enacted by other countries such as the United States. However, it is of equal importance for such laws to be enacted in developing countries. This point was illustrated dramatically at Bhopal in India in December 1984 when an industrial leak of 40 tonnes of methylisocyanate (MIC) caused 3300 deaths and 200,000 injuries (mainly respiratory problems and eye damage). MIC is an intermediary chemical used in the manufacture of plastics, dyes and pesticides and is made at only three locations in the world. The Bhopal plant, subsequently deemed of substandard design by the Indian government, was surrounded by a sprawling slum of 200,000 people, and this local community was totally unprepared for an accident. There were no emergency plans or equipment, and doctors knew little either about how to treat the victims or what long-term effects the gas would cause. The owners of the chemical plant, Union Carbide, paid out $470 million in compensation to victims and their families. Such large financial compensation must not become an acceptable alternative to having effective pollution-control devices and emergency plans at plants producing highly toxic chemicals.

One group of substances widely used in industry since the 1930s but whose toxic nature was not fully recognized until the 1960s was that of polychlorinated biphenyls (PCBs). PCBs are used in epoxy paints, protective coatings for wood, metal and concrete, and carbonless reproducing paper, but their largest use is probably as coolants and insulators in high-voltage transformers. When materials containing PCBs are incinerated, the PCBs vapourize and form a gaseous pollutant which is dispersed throughout the environment, accumulating especially in aquatic organisms. Burning of PCBs may also release dioxin. PCBs have been found in brown seals off the coast of Scotland, shrimps in Florida, cod in the Baltic Sea, and mussels in The Netherlands (Sandbach, 1982). In Britain, relatively high residue levels have been found in freshwater fish-eating birds, predatory terrestrial birds and in certain marine species such as the guillemot and kittiwake. PCBs may have accounted for massive losses of birds in the Irish Sea in 1969. Some workers have claimed that the thin egg shells produced by some predatory birds in recent years are the result of induction of liver enzymes by PCBs and/or organo-chlorine insecticides. They argue that induction may disturb the balance of steroids that regulate calcium and that this can lead to reduced deposition of the element during egg-shell formation. In recognition of the threat to wildlife posed by PCBs, action by the Organization for Economic Co-operation and Development (OECD) in the 1970s, as well as by the United States Toxic Substances Control Act of 1976, attempted to limit the production and use of PCBs (Schweitzer, 1983).

The belated way in which the effects of toxic chemicals are recognized can be illustrated by the example of vinyl chloride. Vinyl chloride is used to make polyvinyl chloride, a plastic which is used for many purposes, including food wrappings, water pipings and bottles for beverages. Minute amounts of vinyl chloride are found as a contaminant of the finished product and pass from it into food and drink. Vinyl chloride itself has been used as a propellant for hair sprays and insecticides. Only after vinyl chloride was widely used was it belatedly discovered that it could cause cancer. Workers exposed to high concentrations of vinyl chloride during its manufacture were found to experience an increased incidence of angiosarcoma, which is a tumour arising in the blood vessels of the liver (Creech and Johnson, 1974; Tabershaw and Gaffey, 1974). Angiosarcoma is an unusual type of cancer so it would therefore have been possible to detect a hazard to the health of the general public if one had been produced for exposures to low concentrations of vinyl chloride. However, had it been one of the more common cancers, such as lung or stomach cancer, then the risk might not have been detected for many years (UK Royal Commission on Environmental Pollution, 1974). This example highlights the fact that inadequate testing of a chemical may result in it gaining widespread use well before its carcinogenic nature is identified.

Although many of the toxic chemicals of current concern are products of modern technology, some substances have existed as pollutants for much longer, even if they have only recently been identified. One such group is that of polycyclic organic matter (Bjorseth, 1983). When a material containing carbon and hydrogen is burned inefficiently, such as in domestic fires and during refuse burning, polycyclic organic hydrocarbons (POM) are produced. Polycyclic aromatic hydrocarbons (PAH), such as benzo (α) pyrene (BαP), are the most significant of these organic materials because many of them are known to be carcinogenic in animals and people (Nikolaou et al., 1984). As early as 1775, a high incidence of scrotal cancer was reported amongst chimney sweeps which was attributed to exposure to soot and which is now attributed to the PAH present in soot and other products of incomplete combustion. Researchers are suggesting that the increased incidence of lung cancer measured in urban populations compared with rural areas is a result of the higher concentrations of PAH in urban areas (Walker, 1975). Fortunately, urban concentrations of PAH are decreasing. In the United States, the annual mean concentration of BαP (employed as an indicator of PAH) decreased from 3 $\mu g/m^3$ in 1966 to less than 1 μg/m^3 in 1975. Similarly, in London BαP declined by a factor of ten between 1935 and 1965.

Having accepted the proliferation of chemicals without much control in the past, societies now face the expensive and complex tasks of identifying those which are dangerous and then deciding what to do about them. During the last ten years, most industrialized countries have passed laws intended to control the manufacture and use of toxic chemicals. However, the implementation of these chemical-control laws has focused on managing new chemicals – that is, those substances which have been introduced to the market since the laws came into effect. This still leaves 50,000 to 70,000 chemicals introduced over the past 50 years, with little or no knowledge of their chronic effects. The task of reviewing all of these chemicals is a daunting but a necessary one. International collaboration is essential and one step in the right direction has been the establishment by the United Nations of the International Register for Potentially Toxic Chemicals in 1978. This register lists information on environmental and toxicological effects of chemicals as well as information on regulatory actions and the legal status of chemicals in various countries. Such a register is vital in order to provide the data-base from which an international treaty on toxic chemicals may be prepared. Even so, for the control of toxic chemicals to be ultimately successful, industrial and commercial organizations, and also the consumer, need to be more sensitive to health and environmental concerns and to reflect this sensitivity in the development, manufacture, use, and handling of chemicals.

3   Odours, Noise and Waste Heat

3.1   Odours

Although some odours (for example, lemon) are considered pleasant and clean and are used for commercial purposes, odours are generally considered unhealthy annoyances to be removed from the air. This is in line with the view held by the World Health Organization's definition of health as 'a state of complete physical, mental, and social well-being and not merely the absence of disease or infirmity' (Engen, 1972). Odours may affect the well-being of people by eliciting unpleasant sensations, by triggering possibly harmful reflexes and other physiologic reactions, and by modifying olfactory function. Unfavourable responses include nausea, vomiting, and headaches; induction of shallow breathing and coughing; upsetting of sleep, stomach and appetite; irritation of eyes, nose and throat; reduction in the enjoyment of home and external environment; distur-bance; annoyance; depression; and sometimes a decrease in heart rate and constriction of blood vessels of the skin and muscles (US Board on Toxicology and Environmental Health Hazards, 1979, 1980).

Odours may arise from any industrial activity, most commonly from food processing, fish processing, meat packing and rendering, pig and poultry farming, the chemical and petroleum industry, sewage and refuse disposal, paint spraying operations, and the manufacture of paints and plastics. Although odours are in some circumstances detectable several kilometres from source, odour pollution is essentially a local problem. Nevertheless, odour pollution is a significant problem with as many as 25 million residents in the USA perceiving community odours as problems and desiring some form of abatement (Copley International Corporation, 1970, 1971; Flesh et al., 1974).

Odours are typically a mixture of complex gases which stimulate the olfactory or smelling sense. Any distinctive smell, be it a sweet odour from a perfumery or a foul odour from a sewage works, may be objectionable when one has to live with it continuously. The most unpleasant odours caused by industrial activities seem to be those which contain sulphur and nitrogen compounds (Kelly, 1979). Odour substances (osmorphorics) which have received attention in odour-pollution control include ammonia (described as pungent), hydrogen sulphide (rotten eggs), trimethylamine (fishy, pungent), phenols (medicinal), as well as mercaptans and dimethylsulphide. Hydrogen sulphide was associated with a major pollution accident at Poza Rica, Mexico, in 1950 when a plant recovering sulphur from natural gas accidentally released a quantity of hydrogen sulphide for about 26 minutes. This resulted in 22 deaths and 320 people being hospitalized with respiratory and central-nervous-system disorders.

People have differing reactions to odours: the response to an odour may be affected by the subject's age, sex, profession, attitudes towards air pollution in general or towards the source in question, and differences in earlier experience of related environmental events. Another complication in assessing odour problems relates to the phenomenon known as odour fatigue. Given sufficient time, a person can become accustomed to almost any odour and be conscious of it only when a change in intensity occurs (Wark and Warner, 1981). The subjective nature of odour pollution has resulted in a variety of measurement procedures being employed. However, the measurements of odours generally fall into two categories: the determination of the threshold concentration of odorous substances, and the establishment of the type and intensity of atmospheric odours.

The intensity of an odour is some numerical or verbal indication of the strength of the odour: for example, a survey of odour pollution in the Swedish city of Utrecht involved 300 residents who were required to classify the air at specified times of the day as either not annoying, slightly annoying, annoying, very annoying or extremely annoying. Another method utilizes reference standards which are prepared from an odorous substance or substances whose smell resembles that produced by the industrial activity under investigation (for example, solvent emissions). The number of standards in the series may range from 4 to 12, depending on the range of odour intensities expected to be encountered and the possibility of being able to differentiate between adjacent members of the series. The lowest standard intensity must be just above the odour threshold, while the highest standard intensity must be above the maximum strength expected to be encountered in the field (Flesh et al., 1974). Odour judges are trained with the reference standards and then taken to an odour-affected area. An example of this method was that employed to determine the intensity patterns of odours caused by the operations of an oil refinery and an animal rendering plant in Philadelphia (figures 3.1 and 3.2).

In addition, the intensity of an odour may be assessed using the Odour Unit. This is a dimensionless number greater than one, which is the number of dilutions with clean air which are required to reduce a sample of the odour-bearing gas down to that level where half the number of persons in an 'Odour Panel' (typically between two and twenty individuals, some of

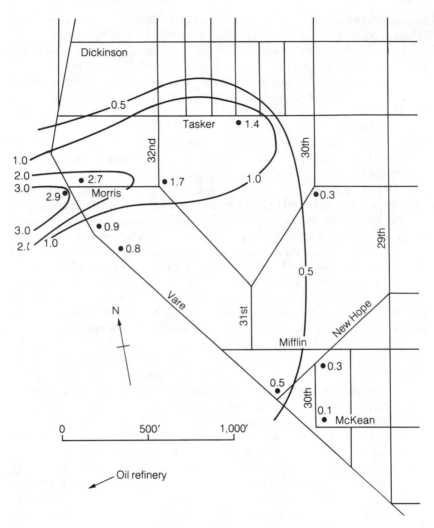

Figure 3.1 Mean odour intensities around an oil refinery and animal rendering plant in Philadelphia
Source: Flesh et al., 1974

whom have received training) could detect it and the other half could not. It is possible to use the dilution method directly in the field using the Scentometer. This instrument allows one to vary the ratio of contaminated air to clean air (dilutions to threshold) and the instrument has a limited number of pre-set dilution ratios so that readings can be quickly obtained. Experience with the Scentometer has shown that odours above 7 D/T (dilutions to threshold) will probably cause complaints, while those above

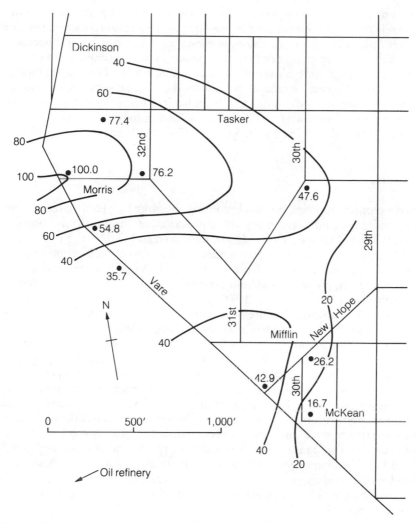

Figure 3.2 Percentage of time odour was detected around an oil refinery and animal rendering plant in Philadelphia
Source: Flesh et al., 1974

31 D/T can be described as a serious nuisance if they persist for any length of time (Leonardos, 1974).

Given recognition of odour-pollution problems, there are two control approaches which are possible. Firstly, one can reduce the concentration of odorous substance so that the smell is less intense and therefore less objectionable. This is achieved by diluting the odour with large quantities of clean air before release, by discharging the odour through taller stacks so as to allow greater dispersal and dilution, by removing the odour before it is released to the atmosphere by adsorption, absorption, or oxidation, or by chemically converting the odorous substance to one which is less odorous. Secondly, one can change or mask the odour so that the unpleasant smell is hidden in other more pleasant and acceptable smells, but this method is controversial because it requires the addition of foreign substances to the air (the opposite of air cleaning) and because its effectiveness has yet to be convincingly demonstrated (US Board on Toxicology and Environmental Health Hazards, 1979, 1980; Kelly, 1979; Wark and Warner, 1981).

3.2   Noise

In developed countries, noise is becoming the most frequently cited irritant in residential and community dissatisfaction of the local environment (Noble and Harnapp, 1981). Noise may adversely affect the health and well-being of individuals or communities. Noise can disturb people's work, rest, sleep, and communications; it can damage hearing and evoke psychological, physiological, and possibly pathological reactions (World Health Organization, 1972, 1980b).

Sound is transmitted as pressure disturbances in the air which vibrate the ear drum to cause the sensation of hearing. A mechanical energy flux accompanies a sound wave, and the rate at which sound energy arrives at, or passes through, a unit area normal to the direction of propagation is known as the sound intensity (normally measured in watts per square metre). Sound intensities of practical interest cover a very wide range, and are therefore measured on a logarithmic scale. The relative intensity level of one sound with respect to another is defined as ten times the logarithm of the ratio of their intensities. Levels defined in this way are expressed in decibels (dB). This means, for example, that if traffic noise registered 70 dB and ordinary speech 60 dB, then the traffic noise is not 10 per cent louder than speech, but rather ten times louder. Thus, a relatively small change in decibel level can mean a tremendous change in the intensity of noise (table 3.1).

In a sound-level meter it is the pressure of the sound waves which is actually measured, but as there is a direct relationship between intensity and pressure, the meter gives a direct reading in decibels. Various filters (A, B, C and D) are employed to weight sound pressure level measurements so as to provide the closest relationship between physical measurements and subjective evaluations of the loudness of noise (World Health Organization, 1980b). The A-filter is so widely adopted that levels measured with this filter, which should be expressed as dB(A), are usually quoted simply as dB in the literature.

Noise levels have increased noticeably during the past few decades as population density has increased, new technological developments have

Table 3.1   Sound intensity level (dB) for selected sources

140	Jet engine (at 25 m)	⎫
130	Rivet gun	⎬ Injurious Range
120	Propeller aircraft (at 50 m)	⎭
110	Rock drill	⎫
100	Metal-working shop/foundry	⎬ Danger Range
80-90	Heavy lorry	⎭
80	Busy street	⎫
60-70	Private car	⎪
60	Ordinary conversation (at 1 m)	⎪
50	Low conversation (at 1 m)	⎬ Safe Range
40	Soft music	⎪
30	Whisper (at 1 m)	⎪
20	Quiet town dwelling	⎪
10	Rustling leaf	⎭

been introduced, and road and air traffic have increased in size and volume (Milne, 1979). Road and air traffic tend to cause the most serious forms of noise pollution to which the public are exposed and this applies in most urban areas, as shown in figure 3.3 for a Chinese city.

As a common measure of environmental noise, and an index to which health criteria may be related, the concept of equivalent continuous sound, L_{eq}, has been introduced. This index is defined as the notional steady sound-level which would, over a given period of time, produce the same sound energy as the actual fluctuating sound. For describing the 24-h general noise environment, a weighted average such as the day-night average sound level (L_{dn}) may be used to take account of sensitive periods of the day or night. In the United States, the Environmental Protection Agency reports that nearly 50 per cent of the nation's population is exposed to L_{dn} (derived by increasing the sound levels between 2200–0700h by 10 dB) in excess of 55 dB and 10–15 per cent to L_{dn} of 70 dB or more. The UK Royal Commission on Environmental Pollution (1974) estimated that half of the UK population could be subjected to traffic noise exceeding the National Advisory Council's limit of acceptability around residential areas (70 dB for 10 per cent of the 18-h day from 0600–2400h). Similarly, it has been estimated that in OECD countries, about 55 per cent of the population (i.e. 400 million people) is exposed to a noise level of over 55 dB and 15 per cent is exposed to noise levels above 65 dB (OECD, 1980b).

Noise disturbance in the home arises from motor cycles, cars and lorries, aircraft, nearby industrial premises and construction sites, barking dogs, pneumatic drills, neighbours' radio and television, and people shouting in the street. This excess noise reduces the quality of the residential environment, reduces property values, imposes a major cost upon society through the need for sound insulation of buildings, and, above all, causes

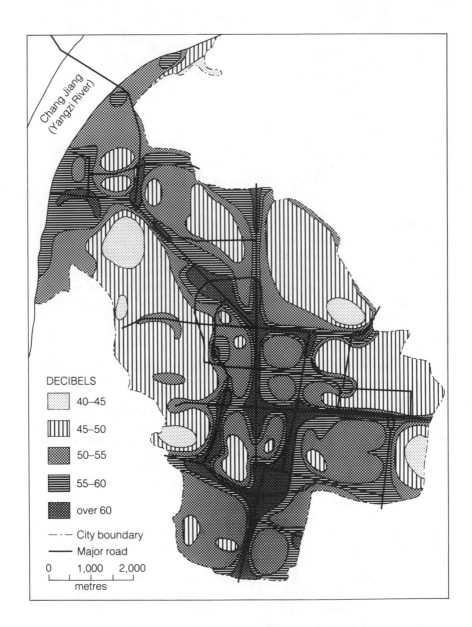

Figure 3.3 Spatial distribution of day-time environmental noise levels in Nanjing, China, clearly reveals the very noisy areas along major roads
Source: Noble, 1980

considerable human stress (Briggs and France, 1982). In addition, at extreme levels wildlife may also experience stress. Risk of permanent hearing loss also increases with noise levels. The risk is negligible for L_{eq} of less than 70 dB for a 24-h period, 75 dB for an 8-h period, 78 dB for a 4-h period, 81 dB for a 2-h period, or 84 dB for a 1-h exposure. Such high noise-levels mean that the risk of permanent hearing loss is an occupational rather than an environmental risk. The problem may be tackled by redesigning or replacing noisy equipment or machinery, by using mufflers and vibration insulators, by constructing noise protection enclosures and by using personal noise protection devices (Beranek, 1971; Mulholland and Attenborough, 1981).

Community dissatisfaction with noise levels relates less to the threat of hearing impairment than it does to general annoyance caused by excessive noise. Noise annoyance may be defined as a feeling of displeasure evoked by a noise and is generally related to the direct effects of noise on various activities, such as interference with conversation, mental concentration, sleep, or recreation. However, there are considerable differences in individual reactions to the same noise due to factors of a social, psychological, or economic nature. At the extreme, noise can induce stress as shown by the increased incidence of high blood-pressure, fatigue and irritability reported in studies of communities in noisy areas such as around airports.

In general, it can be concluded that in residential areas where the daytime noise exposure is below 55 dB L_{eq}, there will be few people seriously annoyed by noise. Although this may be the desirable noise exposure limit for communities, it may be difficult to achieve in many urban areas. Even this level may be considered too high by some residents, especially as substantially lower levels currently prevail in many suburban and rural areas. Given the diversity of the sources of community noise, the problem may be tackled by emission control (for example, lowering noise emission limits on vehicles); by increasing the distance between homes and the noise sources (by careful planning of the location of transport facilities, industrial sites and airports); by the use of protective barriers or partitions alongside highways; by insulating homes (e.g. providing grants for insulation of homes around airports and near motorways); by introducing noise abatement zones; and by educating the public on the adverse effects of excessive noise (World Health Organization, 1972, 1980b).

3.3 Waste Heat

When fuels are used, whether for domestic-space heating, industrial processing or power generation, the sensible and latent heat generated ends up in the atmosphere. Waste heat can be emitted directly to the atmosphere, or indirectly via cooling ponds, rivers, etc. This waste heat acts as an addition to the natural solar energy input of the Earth-atmosphere system and may affect weather and climate. The effects of such artificial heating may be considered at a local, regional and global scale.

3.3.1   Local-scale Effects

The waste heat emitted from single energy-intensive facilities such as power stations, steelworks, and oil refineries may cause localized effects such as increased fog formation and cumulus cloud formation, but it is when many waste-heat emitting facilities are grouped together that significant effects on weather and climate may be observed. The most well-documented example is that of urban areas producing an urban heat-island effect. Urban temperatures frequently exceed those in neighbouring rural areas by on average 1–2°C although during night-time anticyclonic conditions the heat-island intensity may reach between 5 and 10°C.

The excess warmth of urban areas is derived not simply from waste heat released from homes, vehicles and industrial premises but as a result of human alterations of the Earth's surface. In contrast to the soil and vegetated surfaces of rural areas, pavements, roads and buildings readily absorb solar radiation during the day-time and this energy is subsequently released after sunset to the overlying atmosphere. In addition, the reduced plant cover of urban areas leads to a reduction in the amount of solar radiation being utilized in evapotranspiration, thereby increasing the energy absorbed by the urban fabric which in turn warms the urban atmosphere. Artificial heat release can be of more importance than solar radiation absorption in some urban areas, especially during winter: for example, McGoldrick (1980) examined energy use in Greater London for the year 1971 and found that the average daily artificial heat release in the outer suburbs was only about 0–5 $W.m^{-2}$. In the city centre several square kilometres exceeded 100 $W.m^{-2}$, with a maximum for one square kilometre in the inner city of 234 $W.m^{-2}$. These values compare with average daily solar radiation input for the area of 106 $W.m^{-2}$. Even greater energy release occurs in the central part of New York City (Manhattan) where the energy flux density reaches 630 $W.m^{-2}$ (Landsberg, 1981). Oke (1973) has demonstrated a relationship between city size (and building density), as crudely measured by the population total, and the intensity of the urban heat-island. As urban areas grow, so their heat-island intensity increases: for example, the daily maximum temperature and daily minimum temperature in Paris increased by 0.011°C per year and 0.019°C per year respectively between 1891–1968 (Landsberg, 1981).

The consequences of the urban heat-island may be either beneficial or detrimental to the inhabitants: for example, the excess warmth reduces the need for heating in winter but in many cities, this fuel saving is offset by the additional energy needed for air-conditioning in summer. Furthermore, air-conditioning equipment discharges heat to the outside air, thereby adding to the warming of the urban atmosphere. During hot summers, the additional heat stress imposed by the heat island may lead to increased cardiovascular illness, especially in the elderly (Beuchley et al., 1972; Macfarlane, 1978; Tout, 1978).

The excess warmth of urban areas, together with the increased surface roughness of the built-up area, leads to increased precipitation, especially from thunderstorms, in and downwind of urban areas. Thus, for example, analyses from a dense network of meteorological observations in and around St Louis enabled Changnon (1978) to conclude that in the urban area of St Louis and in a fan-shaped area stretching 32–40 km eastwards of the city, there were 'more thunderstorms and hailstorms, more hail and lightning strokes per unit area when they occur, longer lasting periods of thunder activity and hail, and more frequent high winds and damaging hail' than in the surrounding areas. Various thunderstorm characteristics for the study period of 1971–5 summers were increased by a range of 11–116 per cent, hail characteristics in a range of 3–333 per cent, heavy rain by 35–97 per cent, and gusts by 91–100 per cent. The more sizeable percentage changes found for thunderstorms, hail and rainfall were realized in their intensities (Changnon et al., 1977). Changnon (1970), Huff and Changnon (1973), together with Atkinson (1968, 1969, 1970, 1971, 1975) who studied the Greater London area, conclude that the primary causal factor enhancing thunderstorm precipitation varies with a given situation, but that the urban heat-island encouraging convection, and the mechanical effects of the urban fabric causing frictional convergence, are the most important factors responsible for the weather changes. Although Changnon (1978) was unable to discover any urban influence on tornado occurrence in St Louis, studies in Chicago by Fujita (1973) and in Greater London by Elsom and Meaden (1982) suggest that the urban heat-island and increased surface roughness may cause the dissipation and suppression of some tornadoes. The influence of urban factors on tornadoes appears to be restricted to inner parts of metropolitan areas and primarily to weak tornadoes.

Studies by Changnon et al. (1977) of the impact of urban modification on summer weather in and around St Louis suggest that the resulting changes produce greater disadvantages than benefits for the community. Although increased rainfall leads to increased crop yields, this is more than offset by increased flooding, soil erosion and acid rain, together with increased hail damage to crops and property.

Whereas the maximum amount of electric power generated at a single thermal power station is about 3000 MW, there have been recent proposals to build 'power parks', energy parks', or 'energy islands', to generate

Figure 3.4 Large numbers of power stations, such as this 2000 MW coal-fired power station at Didcot in England, may be grouped together to form 'power parks'. The waste heat released may lead to significant changes in local weather

10,000–50,000 MW on a land area of 5–100 km^2 (Williams, 1978). With 33 per cent efficiency, these parks could release 100,000 MW of waste heat within a very small area with the potential to cause major local atmospheric effects. The rate of dissipation of the waste energy into the atmosphere is similar to the dissipation of energy by phenomena such as metropolitan heat-islands, thunderstorms, volcanoes, forest fires and large bushfires (Hanna and Gifford, 1975; Hosler and Landsberg, 1977). It is suggested that power parks will cause deep cumulus clouds to form and lead to increased and enhanced thunderstorm activity (Bhumralkar and Alich, 1976). The enhanced convection over the power parks will induce significant inflow of air at the surface and the resulting concentration of vorticity may lead to increased whirlwind activity with the potential to cause damage to the installations (figure 3.4).

3.3.2   Regional-scale Effects

Where many cities lie in close proximity to one another, such as in the eastern United States, Europe and Japan, their collective waste energy release may have the potential to affect climate at a regional scale since atmospheric circulation might be very sensitive to small perturbations in particular locations (Harrison and McGoldrick, 1981). Such an effect may be analogous to the effect of the open Great Lakes in winter which add approximately 10 TW (10^{13} W) of heat to the atmosphere from an area covering 25 by 10 km; this input is known to intensify low-pressure systems and their associated weather (Barrie et al., 1976). Other analogies include areas of anomalously warm or cool ocean surface temperatures which are known to influence the atmospheric circulation (Perry and Walker, 1977). Thus, for example, it has been shown that anomalies in the ocean surface temperatures in the area off Newfoundland influence atmospheric circulation over western Europe in the following month (Ratcliffe and Murray, 1970), while in the Bay of Biscay a warm sea in July favours warm weather in the following month over Britain (Ratcliffe, 1973). The mechanism by which such anomalies influence atmospheric circulation is apparently through changes in regional thermal gradients which influence the tracks taken by depressions. Currently, anthropogenic waste energy fluxes at the regional scale are much less than the increased sensible and latent heat fluxes from sea surface temperature anomalies. Matthews et al. (1971), for example, estimate waste heat releases over western Europe and the eastern United States (approximately 10^6 km^2) at 0.74 W.m^{-2} and 1.11 W.m^{-2} respectively. With warm sea-surface temperature anomalies producing 40 W.m^{-2}, there is still room for considerable growth of regional waste heat release before comparable effects on the atmosphere may occur.

3.3.3   Global-scale Effects

At present, the total amount of heat released by the sum total of human activities is only approximately 0.01 per cent of the solar energy absorbed at the surface. Total energy consumption will undoubtedly continue to increase and higher energy flux densities will probably become more widely distributed geographically, but it is unlikely that global waste energy release will be sufficient to affect global climate until well into the 21st century (US National Research Council, 1977b).

According to Williams (1978) the total world energy-consumption in 1974 was 7.4 TW, of which the developed countries (population 1160 million) used 6.15 TW (5.33 kW/cap) and the developing countries (population 2760 million) 1.25 TW (0.45 kW/cap). This is equivalent to 0.01 per cent of the total incoming solar energy (Hosler and Landsberg, 1977). If it is assumed that the less-developed countries have a higher growth rate in the future, by the year 2000 the total energy requirement will be approximately 20 TW (developed countries: population 1325 million, per capita energy consumption of 8 kW; developing countries: population 4000–4800 million, per capita energy consumption of 2 kW). For the year 2025 the world energy requirement may reach 30 TW. In contrast, Lovins and Lovins (1982) predict a global energy use by 2030 of 22 TW, based on low growth-rates of energy use. Such values are still small, however, compared with solar energy inputs into the Earth-atmosphere system, and Flohn (1977) suggests that human energy release must reach a minimum of 100 TW before significant global-scale effects on climate will be produced. To reach this figure, with a projected world population of, say, 12,000 million by the year 2075, would require an average energy supply of over 8 kW per capita. This suggests that effects on global climate resulting from total world energy release will not be realized until towards the end of the next century. Even so, some researchers warn that a world population of 20,000 million could ultimately be supported, given high technology, and that it is not inconceivable that the per capita energy consumed by this 'post-industrial society' could reach 40 kW per capita, or four times that of the United States in 1976 (Kellogg, 1978). In the meantime, regardless of what global scenario is most likely to be realized, adverse impacts of waste heat are recognized at the local and regional scales. The adverse effects are important enough to urge the increasing investigation and utilization of energy sources and power production methods that minimize the release of heat to the atmosphere (Revelle and Shapero, 1978).

4 Acid Rain

4.1 Dimensions of the Acid Rain Problem

The acidity of precipitation is determined by the concentration of hydrogen ions (H^+) present and it is commonly expressed in terms of pH values, where pH is defined as the negative logarithm of the hydrogen ion concentration. The pH scale extends from 0 to 14, with a value of 7 indicating a neutral solution. Values less than 7 (distilled water) indicate acid solutions (for example, lemon juice registers pH 2.2, and vinegar pH3); values greater than 7 indicate basic solutions (for example, baking soda registers pH 8.2 and ammonia pH 12). Because the scale is logarithmic, each whole number increment represents a tenfold change in acidity – thus, solutions of pH 6, 5 and 4 contain 1, 10 and 100 micro equivalents of acidity (H^+) per litre respectively (abbreviated μ eq/1) as illustrated in figure 4.1.

'Unpolluted' precipitation is commonly assumed to have a pH value higher than 5.6 (or 5.65). This mild acidity is caused by the presence of carbon dioxide in the atmosphere forming carbonic acid. Natural constituents such as ammonia, soil particles, seaspray, and volcanic emissions of sulphur dioxide and hydrogen sulphide can increase or decrease the pH of precipitation from 5.6. It is not unusual for a pH value of between 4 and 5 to occur at certain locations far removed from human interference (Hibbard, 1982). Similarly, the pH level regularly exceeds 6 in large, dry land areas where there are plenty of basic substances in the material of the soil. Even given this complicating variability in pH levels, researchers have suggested that human activities have caused a dramatic increase in the acidity of precipitation at a local, regional and perhaps even a global one (Likens et al., 1979). Large quantities of oxides of sulphur and nitrogen are being emitted into the atmosphere by the combustion of fossil fuels and industrial processes, and these gases are being converted into strong acids (sulphuric and nitric) which lead to many areas experiencing precipitation of a very low acidity. Precipitation with a pH value lower than 5.6 is termed acid precipitation or simply, acid

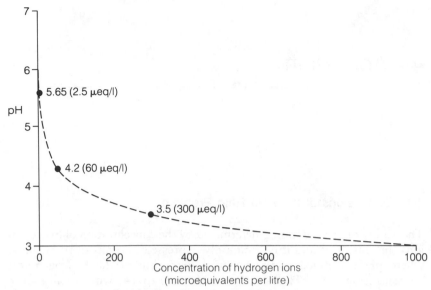

Figure 4.1   The relationship between pH and the concentration of hydrogen ions for aqueous solutions

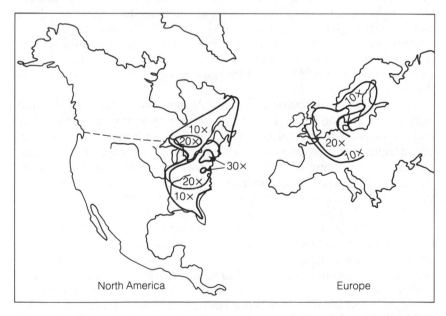

Figure 4.2   Generalized isolines of acidity levels for precipitation relative to the lower limit (2.5 μ eq/1 = pH 5.6) expected for an unpolluted atmosphere. Smaller areas experiencing acid precipitation are not shown. The outer boundary of these maps (1 ×) is vague due to the lack of data and is therefore not shown
Source: Likens and Butler, 1981

rain. The precipitation that fell before the Industrial Revolution and was preserved in glaciers and ice sheets is generally found to have a pH of over 5 and often as high as 6. Currently, extensive areas of Europe and eastern North America are experiencing average pH values of less than 4.5 and even less than 4.0. This represents precipitation acidity up to 30–40 times that which would be expected from an unpolluted atmosphere (figure 4.2). Individual storms may be 500 times more acidic than expected and exceptionally low pH values have been observed during rainstorms as, for example, at Pitlochry in Scotland when on 10 April 1974 a pH value of 2.4 was recorded (Likens et al., 1979).

Although acid precipitation obviously refers to the process of wet deposition of acidic material on the Earth's surface, it also includes the process of dry deposition. In the absence of precipitation, atmospheric pollutants are removed from the atmosphere by gravitational settling and by direct contact with the ground, vegetation and buildings. Typical dry deposition rates of pollutants are 0.1 to 1.0 cm/sec for sulphur dioxide, 0.1 cm/sec for sulphates, 0.2 to 0.5 cm/sec for nitrogen dioxide, and 1.0 cm/sec for nitrates. Dry deposition may make a major contribution to the acidity problem. Both wet and dry deposition are implicitly included under the term acid rain and so would be better termed acid deposition.

Acid deposition is not a new problem: for example, the term acid rain was used first by R. A. Smith in 1872 to describe the polluted air of Manchester, England, which damaged vegetation, bleached the colours of fabrics and corroded metallic surfaces. However, the current concern for acid rain arose when Odén (1968, 1971) analysed the 1956–66 data collected at 160 sites of the European Air Chemistry Network (figure 4.3). He showed that precipitation in parts of Europe had become increasingly acidic over this period. In the late 1950s, the Low Countries (Belgium, Luxembourg, and The Netherlands) were experiencing precipitation with an acidity of pH 4.0 to pH 4.5. By the mid-1960s, not only was this area experiencing acidities less than pH 4.0 but the area of acid precipitation with values below pH 4.5 had expanded to include most of East and West Germany, northern France, eastern England and southern Scandinavia. These findings, given international prominence at the United Nations Conference on the Human Environment in Stockholm in 1972 (Engstrom, 1972), stimulated further national and international research into the problem of acid rain, including the OECD study and the 8-year SNSF project in Norway (OECD, 1977; Overrein et al., 1980).

Following recognition of the existence of widespread acid precipitation in Europe, the probability that a similar problem existed in northeastern USA and adjacent parts of Canada was highlighted by Likens and his colleagues (Cogbill and Likens, 1974; Likens and Bormann, 1974; Likens et al., 1972). Their conclusions were based primarily on records of precipitation chemistry taken in north central New Hampshire (the

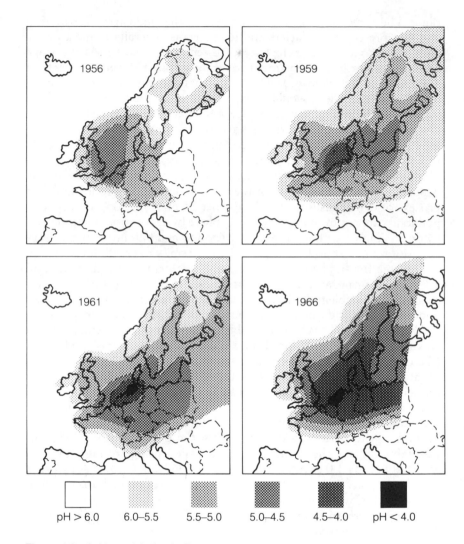

pH > 6.0	6.0–5.5	5.5–5.0	5.0–4.5	4.5–4.0	pH < 4.0

Figure 4.3 Acid precipitation in Europe
Source: Odén, 1971

Hubbard Brook Experimental Forest Network) and New York, as well as on some scattered observations from other locations. From these sources of varying accuracy and length of record, Cogbill and Likens (1974) and Likens et al. (1979) produced isopleths of precipitation acidity for selected periods which revealed increasing acidity between 1955–6 and 1975–6 and an enlargement of the area affected by acid precipitation in eastern North America.

Although isopleths showing the spread of acid precipitation in both Europe and eastern North America have been cited frequently in order to highlight the seriousness of the problem, many researchers have criticized their use. Kallend et al. (1983) have argued that maps based on the European Air Chemistry Network may be misleading in several respects, such as in failing adequately to show the pattern of time variations over the years or the large degree of uncertainty attaching to individual contours, and assuming a geographical homogeneity which is not borne out by detailed calculations with data taken from adjacent sites. When Kallend et al. (1983) analysed the 120 sites individually with five or more years of data between 1956–76, only 29 showed a significant trend of increasing annual average rainfall acidity and five showed a decrease. Critics of the data-base used to define trends in acidity in North America highlight similar problems. Of the ten common sites available for a comparison of 1955–6 and 1965–6, only four showed worsening acidity, two showed improvement, and the other four remained the same. Only two common sites were available for a comparison between 1955–6 and 1972–3, with one showing increasing pH, and the other showing a decrease. If only the common sites were compared, justification for the claim of widespread increasing acidity over time is not supported (Record et al., 1982). Stensland and Semonin (1982) even claim that severe drought and duststorms affected much of the USA in the 1950s, increasing the calcium and magnesium levels in soils, thereby partly explaining differences in precipitation acidity between the 1950s and 1970s.

The need for accurate and long-term information concerning the composition of atmospheric wet and dry deposition has resulted in the establishment in Europe and North America of sampling networks aimed at documenting any further changes in the distribution and composition of acid precipitation. The Canadian Network for Sampling Precipitation (CANSAP) was founded in 1977 while the National Atmospheric Deposition Program (NADP) of the United States commenced a year later (figure 4.4; Likens and Butler, 1981). Since 1976, the United Nations Economic Commission for Euope has assumed responsibility for the monitoring programme in Western Europe and subsequently extended this to Eastern Europe. The European network is called EMEP or the Co-operative Programme for Monitoring and Evaluation of Long-range Transmission of Air Pollutants in Europe.

Very long-term records of lake acidification can provide quite dramatic evidence for the recent accentuation of the acid-precipitation problem. Renberg and Hellberg (1982), for example, analysed diatom assemblages in lake sediment cores in southwest Sweden and found that since deglaciation of the area about 12,500 years ago, the pH of the lakes had decreased from about 7 to about 6 due to natural long-term oligotrophication and acidification. However, during the last few decades

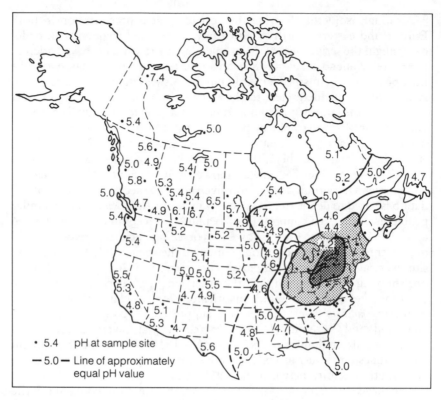

Figure 4.4　Acid precipitation (wet deposition only) in North America in 1982
Source: Miller, 1984

a further, more marked, decrease occurred from values around pH 6 to
the present-day values of about pH 4.5 (figure 4.5). Diatom analyses of
long-term sediment cores from two lakes in southwest Scotland also
revealed marked increases in acidity, especially during the past few decades
(Flower and Battarbee, 1983).

　The principal precursors of acid precipitation are the anthropogenic and
natural emissions of oxides of sulphur and nitrogen which are subsequently
catalytically or photochemically oxidized to sulphuric and nitric acid,
respectively, in the atmosphere (figure 4.6). The acids dissolved in water
appear largely in the form of ions (charged atoms). Anions are negatively
charged ions while cations are positively charged. The ions of concern
are the sulphate ion (chemically abbreviated as $SO_4^=$ or SO_4^{2-}), nitrate
ion (NO_3^-) and hydrogen ion (H^+). Nitric acid releases one hydrogen ion
per molecule, whereas sulphuric acid releases two. Some of the acid is
neutralized by such substances as ammonia (NH_3), derived from natural
and biological processes, whereupon ammonium ions (NH_4^+) are

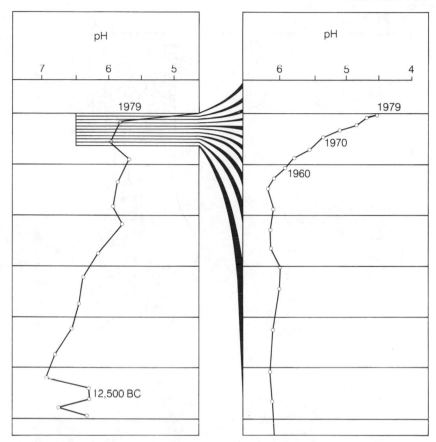

Figure 4.5 The pH of Lake Gårdsjön, Sweden, over a time-scale from 12 500 BC to 1979 AD
Source: modified from Renberg and Hellberg, 1982

formed (Galloway et al., 1976). The various ions are removed from the atmosphere by the two wet deposition processes of rainout and washout. Rainout occurs when material is incorporated into cloud water-droplets or ice crystals which eventually grow to sufficient size to overcome gravity and fall to the ground. Washout occurs when material below the cloud is swept out by rain or snow as it falls.

Fossil-fuel combustion (coal and oil) and industrial processes (principally primary metal production) are the largest contributing human-derived sources of sulphur dioxide emissions. Global anthropogenic sulphur dioxide emission is estimated to be approximately 75–100 million tonnes a year (Swedish Ministry of Agriculture, 1982). Although anthropogenic emissions of sulphur dioxide account for only half of the total global

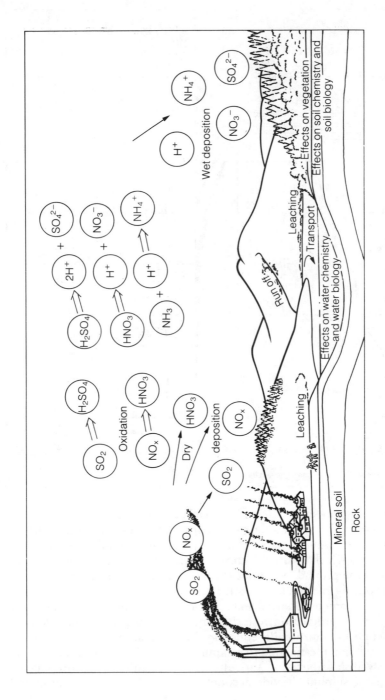

Figure 4.6 Oxides of sulphur and of nitrogen – transport, chemical conversions, deposition, and environmental effects

Source: Swedish Ministry of Agriculture, 1982

emission, they tend to be very concentrated. In areas such as Europe and eastern North America, they account for 90 per cent of total sulphur dioxide emissions. Natural sources of sulphur oxides include seaspray containing sulphate from oceans, organic compounds from bacterial decomposition of organic matter, reduction of sulphate in oxygen-depleted waters and soils, volcanoes, and forest fires (Record et al., 1982). Global anthropogenic emission of oxides of nitrogen for 1978 were estimated at 75 million tonnes (Svensson and Soderlund, 1975). Fossil fuel combustion (coal, oil and natural gas) and transportation (petrol and diesel fuels) contribute most of the anthropogenic oxides of nitrogen emissions. The principal sources of natural oxides of nitrogen appear to be chemical decomposition of nitrates and lightning. Estimated ratios of natural to anthropogenic emissions of oxides of nitrogen range from 15:1 to 1:1 because of the variability in estimates of the natural emissions.

The relative importance of the component acids in precipitation vary spatially and temporarily, but in general similar values apply in eastern North America and Europe. For northeastern USA, measurements suggest that the contributions to total acidity are made up of 62 per cent from sulphuric acid, 32 per cent from nitric acid, and 6 per cent from hydrochloric acid (Record et al., 1982). For Scandinavia the figures are 70 per cent sulphuric acid and 30 per cent nitric acid (Swedish Ministry of Agriculture, 1982), which are similar to the values from Scotland of 71 per cent sulphuric acid and 29 per cent nitric acid (Fowler et al., 1982). However, in some regions, such as the west coast of the United States and in Japan, nitric acid contribution is of greater relative importance. With emissions of oxides of nitrogen increasing at a faster rate than sulphur dioxide emissions, nitric acid will increase in relative importance in its contribution to acid precipitation (Lewis and Grant, 1980). Indeed, this has already been observed in Scandinavia and North America: for example, Cogbill and Likens (1974) suggest that the proportion contributed by nitrates has increased from 22 per cent in the mid-1950s to its present value of over 30 per cent.

Areas that emit most acid precursors are among those which suffer the most acid deposition, but only long-range transport of oxides of sulphur and nitrogen from these sources can explain acid rain experienced at locations remote from significant pollution sources. Similarly, in areas where local emissions have declined but the acidity of rainfall has increased, this points to long-distance transport of pollutants into the area. Although a substantial part of the emissions of sulphur dioxide is deposited near their sources, a significant proportion is dispersed further afield. That proportion of sulphur dioxide not deposited locally will be diffused in the atmosphere and, through oxidation, will be transformed into sulphates. Conversion rates of sulphur dioxide to sulphate range from 0.1 to 10 per cent per hour, depending upon the presence or absence of other pollutants (reactive hydrocarbons and photochemical oxidants) and atmospheric conditions (relative humidity, temperature, sunlight). Sulphates are less subject to dry deposition than sulphur dioxide and consequently remain in the atmosphere for days, travelling up to several thousand kilometres in favourable meteorological and synoptic conditions. Eventually the sulphates are wet deposited by rainout and washout processes. A similar situation applies to emissions of oxides of nitrogen, although gaseous oxides of nitrogen have a low dry-deposition rate, thereby allowing greater proportions to be converted to nitrate aerosols which are finally removed by rainout and washout.

Since mountainous areas enhance and induce precipitation, it is these locations which cleanse the atmosphere of its long-range transported pollutants and thus experience significant acid precipitation. Many researchers point to some pollution-control technologies such as the building of tall stacks for power stations and large industry, in order to reduce local pollution, as having increased the long-distance transport of pollutants and so having accentuated the acid rain problem (figure 4.7; Patrick et al., 1981). Thus, for example, the copper-nickel smelter in Sudbury, Ontario, through which is annually emitted 1.35 million tonnes of sulphur dioxide (in 1973) or 1–2 per cent of the annual world anthropogenic output of sulphur, has a stack over 400 m high (Barnes, 1979).

Although it is clear that mountainous areas with few nearby notable pollution sources are receiving their pollution burden from regions or countries several hundreds or thousands of kilometres distant, it is difficult to determine precisely from which source or sources the pollution originated. Thus, for example, the existence of the acid rain problem in Scandinavia initially pointed to emission sources in the United Kingdom as the cause. That the acid deposition problem became accentuated after the 1950s, coincident with the introduction of the 'tall-stack policy' for United Kingdom power stations, also apparently indicated where the cause lay. However, although southern Scandinavia is indeed subject to frequent airstreams from Britain, those airstreams are only mildly polluted compared with the occasional airstreams which originate from central and eastern Europe (Czechoslovakia, Hungary, Poland, East and West Germany). These latter occasions give rise to highly acidic precipitation episodes (sometimes producing grey or black snows resulting from the presence of lignite ash) which are major contributors to the total annual amount of wet deposition of sulphate in Scandinavia (Smith and Hunt, 1978). Further, Kallend et al. (1983) reveal that of those European Air Chemistry Network stations showing a significant trend of increasing annual average precipitation acidity, this was primarily due to an increased number of intermittent high acidity episodes rather than a substantial increase in general precipitation acidity. Although such evidence may shift

Figure 4.7   Tall industrial stacks encourage the transport of sulphates and nitrates over long distances leading to regional hazes and acid rain

the relative contribution each country makes to another's acid rain problem, the international dimension of this atmospheric pollution problem remains the same. A successful resolution to the problem will require effective action at an international level (refer to chapter 11).

The impact of acidic precipitation on aquatic and terrestrial ecosystems depends not simply on the pH value of the precipitation but also on the capability of lakes and soils to neutralize or buffer the acidic inputs of the precipitation. This ability to tie up the excess hydrogen ions introduced by acid precipitation is largely determined by the composition of the bedrock on which the lakes and soils form, with hard, impervious igneous or metamorphic bedrock giving rise to low calcium and magnesium content producing the most susceptible situation. This explains why parts of southern Scandinavia and eastern North America have figured prominently in highlighting the damaging effects of acid deposition on ecosystems. Considering the sensitivity of lakes and soils in other parts of the world and expected rapid industrialization in the coming decades, countries such as Brazil, Nigeria, southern Africa, India, Malaysia and China are among those areas likely to be exposed to an increased risk of acid deposition effects in the future (Swedish Ministry of Agriculture, 1982). Already, Harte (1983) has reported a pH value of 2.25 in a remote small city in central China, which is one of the most acidic precipitation samples ever recorded.

4.2   Effects on Aquatic Ecosystems

Most concern for the impact of acid rain on aquatic ecosystems is focused on the effects on the fish population. It is generally accepted that increasing lake acidification due to acid deposition has caused fish kills and stock depletion (figure 4.8). Einbender et al. (1982) report the existence of at

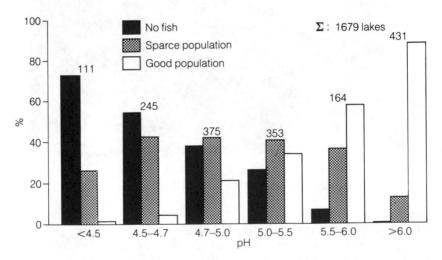

Figure 4.8   Fish status for 1679 lakes in the Norwegian counties of Rogaland, Vest-Agder, Aust-Agder and Telemark grouped according to water pH
Source: Barnes, 1979

least 212 fishless lakes in the Adirondacks, the loss of all fish in 140
Canadian lakes (and the threatened extinction in 48,000 other lakes),
damage to 15 per cent of lakes in Minnesota, the threatened extinction
of trout in half of West Virginia's trout streams, the near total depletion
of buffering capacity in 3000 lakes in the eastern United States, the
acidification of more than 10,000 lakes in Sweden to pH 6 or below (with
5000 lakes having pH values below 5), and reduced populations of salmon
and trout in major rivers in southern Norway. Species of fish vary in their
tolerance to low pH: among the salmonids, the rainbow trout is the most
sensitive, followed by salmon, with brown and brook trout being the least
sensitive (table 4.1).

Adverse effects on fish have been attributed to either sudden, short-term
or gradual long-term decreases in pH. Short-term pH changes, following
the first heavy autumn rains, or especially in the early spring when
snowmelt releases acidic constituents accumulated during the winter, result
in 'acid shock' which leads to fish mortality. The Norwegian SNSF project

Table 4.1   Summary of effects of pH changes on fish

pH	Effects
11.5-11.0	Lethal to all fish
11.5-10.5	Lethal to salmonids; lethal to carp, tench, goldfish, pike if prolonged
10.5-10.0	Roach, salmonids survive short periods, but lethal if prolonged
10.0-9.5	Slowly lethal to salmonids
9.5-9.0	Harmful to salmonids, perch if persistent
9.0-6.5	Harmless to most fish
6.5 6.0	Significant reductions in egg hatchability and growth in brook trout under continued exposure
6.0-5.0	Rainbow trout do not occur; small populations of relatively few fish species found; fathead minnow spawning reduced. Molluscs rare
	Declines in a salmonid fishery can be expected
	High aluminum concentrations may be present in certain waters causing fish toxicity
5.0-4.5	Harmful to salmonid eggs and fry; harmful to common carp
4.5-4.0	Harmful to salmonids, tench, bream, roach, goldfish, common carp; resistance increases with age; pike can breed, but perch, bream, and roach cannot
4.0-3.5	Lethal to salmonids; roach, tench, perch, pike survive
3.5-3.0	Toxic to most fish; some plants and invertebrates survive

Source: Schofield, 1976

Figure 4.9 Many thousands of lakes in Scandinavia have been depleted of their fish populations. Lakes develop an extensive layer of Sphagnum moss which covers the lake bottom and creates a new lake shore
Source: R. A. Barnes

found that 80 per cent of the pollutant content of the winter snowpack is released in the first 30 per cent of the meltwater (Overrein et al., 1980). Gradual decreases in pH with time lead to prolonged acidity which interferes with fish reproduction and spawning, so that over time, there is a decrease in fish population and a shift to a smaller number of more tolerant older and larger fish. An additional problem is that acid deposition leads to the mobilization of toxic metals, especially aluminium, and this can be another factor contributing to fish mortality. The aluminium obstructs the ability of the fish to take in salt through their gills necessary to maintain their asmotic balance, and by coagulating the mucous coatings of the fish's gills, this interferes with its ability to breathe (Gorham, 1982).

It is not simply fish species which have suffered from lake acidification. Observations of lakes in southern Scandinavia and eastern North America have shown that where the pH level has fallen, a decline in the lower phytoplankton biomass and species variety takes place. Hendrey et al. (1976) have shown that where lake pH values were below 5, about 12 species were observed, falling to as few as three in some very acidified lakes. In contrast, where pH values were over 6, between thirty and eighty species were present, and the phytoplankton biomass was between three and nine times greater. Such reduction in lake diversification reduces the variety of food available through the trophic levels. Particularly noticeable in some Swedish lakes, with pH below 6, is the expansion of the moss Sphagnum, which covers the lake bottom, displacing other plant life and further increasing lake acidity by removing base materials such as calcium from solution and releasing hydrogen (figure 4.9). Epiphytic and epilithic algae colonies also seem to be expanding in the acidic lakes and watercourses of southern Scandinavia.

4.3   Effects on Terrestrial Ecosystems

Acid deposition may cause damage to terrestrial ecosystems by increasing soil acidity, decreasing nutrient availability, mobilizing toxic metals, leaching important soil chemicals, and changing species composition and decomposer micro-organisms in soils. Much concern has been expressed that acid precipitation causes a reduction in forest productivity. Vast areas of forests, especially at altitudes above about 600 m, have recently been damaged in Europe with a significant proportion having been severely damaged (estimated at 5 per cent in 1985) or even killed (0.2 per cent in 1985). In West Germany, four of the nation's most important tree species, Norway spruce, white fir, Scotch pine, and beech have shown alarming signs of deterioration (table 4.2; Wetstone and Foster, 1983). Different trees are affected at different rates. On spruce, top growth slows, resulting in a 'stork's nest' effect. Needles begin to yellow, usually on the

Table 4.2   Percentage of forested area affected by air pollution in West Germany since 1982

Extent of damage	1982	1983	1984	1985	1986
Light damage (20% of needles or foliage lost)	6	24.7	32.9	32.7	34.8
Medium damage (20–60% of needles or foliage lost)	1.5	8.7	15.8	17.0	17.3
Severe damage/dead (60% of needles or foliage lost)[b]	0.5	1.0	1.5	2.2	1.6
TOTAL	8.0	34.4[a]	50.2	51.9	53.7

[a]Substantial increase on 1982 figures due to the establishment of a more thorough investigation procedure.
[b]Given that trees are usually cropped before death to prevent fungal attack and, in particular, bark beetle from destroying the commercial value of the wood, this leads to an underestimation in this category.
Source: Information supplied by the Ministry of Interior, Bonn

upper side of branches, and then fall off. Growth appears stunted and side branches die. In deciduous trees, leaf discolouration, early leaf fall, death of tree tops, damage to bark as well as lack of natural rejuvenation, are observed.

In eastern Europe, forest problems are even more severe than central Europe (Tomlinson and Silversides, 1982). Nearly 700,000 hectares (or 16 per cent) of forests in Czechoslovakia have been damaged with deaths in widespread areas of the mountainous Krokonose Park on the Polish-Czech border, where precipitation reportedly averages pH 3.8 (over 60 times more acidic than average unpolluted precipitation), and in Krusne Park on the East German-Czech border. Not surprisingly, a similar situation applies in Poland. The problems in eastern Europe are particularly severe because of the vast emissions of sulphur dioxide from (largely uncontrolled) industrial plants and the high sulphur content (3.5 to 14 per cent) of brown coal or lignite burned as compared with bituminous coal (1 to 2 per cent).

In a 15-year German study, Ulrich et al. (1980) and Ulrich and Pankrath (1983) showed that acid rain is leaching important plant nutrients such as calcium, magnesium and potassium from the soils, making them unavailable to trees. In addition, acid rain mobilizes the aluminium in forest soils (from harmless soil compounds such as aluminium silicate) which decreases the ratio of calcium to aluminium in soil solutions to the extent that root growth is impaired. The fine roots of trees are damaged by the toxicity of aluminium, which together with nutrient-deficiency, causes a stress condition of crown die-back (leaves or needles at the tree-top turn yellow, then brown, and eventually drop off) and the eventual death of the tree. Seedlings may not survive because of extensive damage to their roots. However, the Ulrich hypothesis does not explain why trees growing in a wide range of soil types, even calcareous soils, are apparently suffering acidic deposition effects.

It may be that there are a variety of reasons why an increasing number of European trees are becoming damaged or killed. There is wide agreement that the recent forest damage can predominantly be ascribed to atmospheric pollution even if there remains uncertainty concerning which pollutants are to blame at which locations and what the response mechanisms are for triggering damage, as well as over how the contributing natural stress factors should be weighted (Hinrichsen, 1986). Various researchers point to the potential phytotoxic effect on trees of high concentrations of sulphur dioxide around industrial areas; to the higher acidities (e.g. pH of 3.5) of occult precipitation (mist, fog) from clouds enveloping high-altitude forests for considerable portions of the year; to the increasing concentrations of photochemical oxidants (PAN, ozone) in Europe; to the increase in oxides of nitrogen and hydrocarbons from motor vehicles; and to synergistic effects between many of these forms of pollution. Whatever the reason, it is recognized that stressed trees then become vulnerable to further damage through drought, wind damage (through weakened root systems), snow damage (the brittle tops of trees snap beneath the weight of snow), insect (e.g. the bark beetle), fungal and virus attack.

Forest damage similar to that observed in parts of Europe is being observed in North American areas subject to acid rain such as on mountain tops in the southern Appalachians, the Adirondacks, Vermont and New Hampshire where die-back of red spruce is evident (Einbender et al., 1982; Johnson and Siccama, 1983; United Nations Economic Commission for Europe, 1985). Explaining the widespread damage to forests in Europe is made more difficult by the existence of only limited acid rain effects on the extensive forests of southern Scandinavia. However, acid deposition rates in Scandinavia are considerably less than those observed in central and eastern Europe so obvious tree damage such as reduced foliage and brittle crowns may not be expected. Further, such negative effects as leaching of nutrients (calcium, magnesium, potassium and phosphorous) may be temporarily offset in established forests by the increasing fertilizing effect of nitrates. However, this beneficial aspect of acid rain may be short-lived because over the long-term, forest productivity may be limited by the lack of nutrients other than nitrogen. To compensate for decreasing soil mineral reserves, the tree's overall root mat system eventually shrinks, making it susceptible to drought in summer and wind-throw in winter. Overfertiliz-ation by nitrogen also makes trees more palatable for insects. Gorham (1982) even suggests that the acidity may hinder the ability of the forest's bacteria and fungi to recycle nitrogen from decaying plants into the soil, robbing the trees of as much nitrogen as the precipitation deposits.

Figure 4.10 Lead was used to 'point' blocks of building stone in earlier centuries. Acid deposition has eroded the stonework of St Paul's Cathedral, London, such that the lead now stands a centimetre proud of the stonework

Source: London Scientific Services

In addition to forests, attention has been focused on acid precipitation effects on a wide range of crops. Record et al. (1982) report on an American study which examined thirty two commercial crops grown under controlled environmental conditions and exposed to simulated (sulphuric) acid rain of pH 3.0, 3.5, 4.0 and 5.7. Preliminary results indicate that some crops suffered extensive foliage damage and yield reduction but others sustained little apparent injury, even under these severe exposure conditions. A similar variability in response to different rain acidities was reported by Hibbard (1982) when of twenty seven crops tested, five were inhibited, six stimulated, and sixteen experienced no observable effects.

4.4   Effects on Human Health

Heavy-metal cations, such as cadmium, nickel, lead, manganese and mercury, may be bound to colloidal particles in the soil. Excessive amounts of hydrogen cations introduced into the soil by acid precipitation may exchange for these heavy-metal cations, thereby releasing the metals into the soil and into watercourses (Babich et al., 1980). The metallic compounds may contaminate edible fish and drinking water supplies, and thereby be passed on to people. At some locations, concentrations of these elements in drinking water exceed the national drinking water standards, while in some lakes high methyl-mercury content in fish could have potentially adverse effects. Water supplies become contaminated because increasingly acidic water leaches copper and lead from water supply systems. Another concern is that acid deposition may accelerate the leaching, mobilization and enhanced biological availability of toxic heavy metals and other toxic chemicals in past, present and future hazardous waste (and other solid waste) landfills (Einbender et al., 1982).

4.5   Effects on Materials and Visibility

Acid precipitation can accelerate corrosion of metals and erosion of stone (figure 4.10). However, because sulphur compounds are responsible for most of these effects it is very difficult to distinguish between the effects of pollution originating from local as opposed to distant sources (refer to chapter 2, section 2.1).

The transport of sulphates over long distances can lead to reductions in visibility over extensive areas and at locations distant from pollution sources. Dense summer acid hazes are now experienced over extensive areas of rural southern Britain, being closely related to the advection of polluted continental air masses (Barnes and Lee, 1978; Lee, 1983a, 1983b). Similarly, hazes mapped by Hidy et al. (1978) and observed on satellite images by Lyons and Dooley (1978) covered vast areas of the northeastern USA and were associated with polluted air mass advection from the Mid-West. During winter and early spring, air masses entering polar regions are charged with sulphates which are occasionally sufficient to reduce visibility and cause 'Arctic haze'. Sources for the sulphates are the distant industrial regions of Europe, North America and eastern Asia, with most researchers pointing to Europe and eastern USSR as the major contributors (Barrie, 1986; Shaw, 1982). Not only does the increased frequency of pollution hazes in rural and wilderness areas cause dissatisfaction and loss of amenity to many, but they also have the potential for affecting global climate (refer to chapter 2, section 2.1).

5 Ionizing Radiation (Radionuclides)

5.1 Introduction

Radioactive materials are present in the human environment as the result of natural processes and of human technological developments. These materials produce 'ionizing radiation' such as X-rays, gamma-rays, alpha-particles, beta-particles, electrons, protons, neutrons and cosmic rays which ionize the atoms of substances which they penetrate. Radiation such as ultraviolet and visible light, infra-red radiation and radio waves do not produce ionizations. Ionization is the process by which a fast-moving quantity of energy is transferred to some of the atoms of the material through which it is travelling, leaving them as electrically charged ions. Although the physico-chemical changes caused by the ionization of the atoms of living matter occur in a fraction of a second, the processes whereby these changes eventually lead to such changes in living matter as cell death, malignant tumours (cancer), and genetic mutations may take hours, months, or even decades (Coggle, 1983).

With few exceptions, radioactive materials released into the environment become involved in a complex series of physical, chemical, and biological processes. Some of the processes lead to progressive dilution, others to physical or biological reconcentration, followed by transfer through various or sometimes interdependent pathways to people (ICRP-29, 1979). Simplified representations of some of these pathways are shown in figures 5.1 and 5.2. Some compartments in the model could be subdivided: for example, radionuclides are transferred from animals to people by drinking milk or by eating the animal meat, and while the consumption of milk is a fast pathway, that of meat is slower. The radiation dose received by people is the sum of the external irradiation (from deposited activity and atmospheric activity) and internal irradiation (following inhalation and ingestion of radioactive substances). The ionizing radiations to which people are subject are of varying types and although they interact with living matter in a similar way, different types of radiation differ in their effectiveness in damaging a biological system. Thus, for example, 10 grays

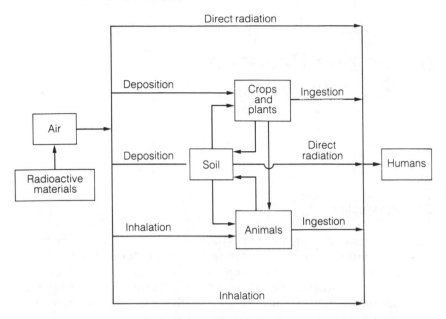

Figure 5.1   Simplified pathways between radioactive materials released to the atmosphere and to humankind
Source: ICRP-29, 1979

(Gy) of alpha-particles will cause the same damage in living tissue as 200 Gy or 200 sievert (Sv) of X-rays, gamma-rays, or electrons (table 5.1 explains the units employed). This is in spite of alpha-particles being less penetrating than gamma-rays. The considerably higher relative biological effectiveness or quality factor (Q) of alpha-particles is believed to be due to the fact that the dose is delivered over short tracks in tissue along which ionization is dense. Figure 5.3 provides details of the quality factor for some radiations. To obtain the biologically effective dose for a mixture of radiations, the dose equivalent (units = sievert) is the sum of doses of each type of radiation (in grays) multiplied by the quality factor (Q) of each radiation type.

Radioactive contamination is removed through the disintegration of an isotope through a series of daughter elements until a stable element is produced. The radioactive decay, through which a radionuclide loses half its radioactivity, is a fixed length of time for each element, and is referred to as the half-life of a substance. As a rule of thumb, the radioactivity of an element has usually disappeared (has reached 0.1 per cent of the original value) after an equivalent time-period of ten times its half-life (Bach, 1972). Half-lives of various radionuclides, together with the principal types of radiation they emit, are summarized in table 5.2.

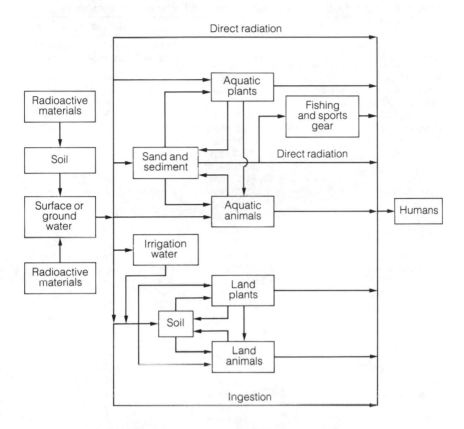

Figure 5.2   Simplified pathways between radioactive materials released to the ground or surface waters (including oceans) and to humankind
Source: ICRP-29, 1979

Table 5.1   Current and past units employed in the measurement of ionizing radiation

Quantity	New named unit and symbol	In other SI units	Old special unit and symbol	Conversion factor
Exposure	–	$C\ kg^{-1}$	röntgen or roentgen (R)	$1\ C\ kg^{-1} \sim 3876\ R$
Absorbed dose	gray (Gy)	$J\ kg^{-1}$	rad (rad)	$1\ Gy = 100\ rad$
Dose equivalent	sievert (Sv)	$J\ kg^{-1}$	rem (rem)	$1\ Sv = 100\ rem$
Activity	becquerel (Bq)	s^{-1}	curie (Ci)	$1\ Bq \sim 3.7 \times 10^{-10} Ci$

Table 5.2   Selected radionuclides – their half-lives and principal radiations emitted

| Radionuclide[a] | Half-life | Principal radiations | | |
		alpha-particles	beta-particles	gamma-rays
Argon-41	1.8 hours		✓	✓
Barium-140	12.8 days		✓	✓
Caesium-134	2.1 years		✓	✓
Caesium-137	30.1 years		✓	
Caesium-139	9.5 months		✓	✓
Carbon-14	5.7×10^3 years		✓	
Cerium-141	33.0 days		✓	✓
Cerium-144	284.0 days		✓	✓
Iodine-129	1.6×10^7 years		✓	✓
Iodine-131	8.1 days		✓	✓
Iodine-132	2.3 hours		✓	✓
Krypton-85	10.7 years		✓	✓
Molybdenum-99	1.1 days		✓	✓
Plutonium-237	46.0 days			✓
Plutonium-238	87.8 years	✓		
Plutonium-239	2.4×10^4 years	✓		
Plutonium-240	6.5×10^3 years		✓	
Polonium-210	138.0 days	✓		
Potassium-40	1.3×10^9 years		✓	✓
Radium-224	3.6 days	✓		✓
Radium-226	1.6×10^3 years	✓		✓
Radium-228	5.8 years		✓	
Radon-222	3.8 days	✓		
Ruthenium-103	39.5 days		✓	✓
Ruthenium-106	1.0 years		✓	
Strontium-89	50.5 days		✓	
Strontium-90	28.5 years		✓	
Tellurium-132	1.3 hours		✓	✓
Uranium-233	1.6×10^5 years	✓		
Uranium-234	2.5×10^5 years	✓		
Uranium-235	7.1×10^8 years	✓		✓
Uranium-236	2.4×10^7 years	✓		
Uranium-238	4.5×10^9 years	✓		
Uranium-239	23.5 months		✓	✓
Xenon-133	5.3 days		✓	✓
Xenon-135	9.2 hours		✓	✓

[a]Isotopes are characterized by a number which represents the total number of particles (protons plus neutrons) in their nuclei, e.g. uranium-235 has 92 protons and 143 neutrons whereas uranium-238 has 92 protons and 146 neutrons.

5.2   Effects on People

Evidence for the effect of ionizing radiation on people is derived from several sources. The survivors of the atomic bomb detonations at

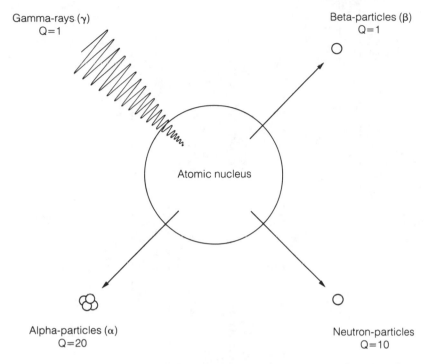

Figure 5.3 Principal types of ionizing radiation and associated quality factor (Q)

Hiroshima and Nagasaki in 1945 provide evidence on the effect of substantially uniform 'whole-body' irradiation on people of all ages, both in terms of acute exposure to large radiation doses (table 5.3) and in the long-term effects of lower levels of radiation. Thus, for example, it has been found that levels of received radiation doses correlate significantly with radiation-induced disorders and diseases such as abnormalities in cell-development, chromosomal damage, leukaemia, cancer of the thyroid, lung and breast, impairment of growth and development, and increase in foetal, infant and general mortality (Bach, 1972). Recent research suggests that the radiation effects induced in these Japanese people may have been caused by only half as much radiation as has hitherto been believed (Hawkes et al., 1986).

Evidence on the risk for one organ, the thyroid gland, can also be derived from the occurrence of cancers on this gland in populations of two Pacific Islands, who were exposed to irradiation and radioiodine concentration in the gland from fallout from a weapon test in 1954. Occupational exposure to radiation of both uranium miners and others serving the nuclear fuel cycle, and radiotherapy medical staff, provides additional evidence. Increased lung-cancer fatalities in uranium miners who inhale

Table 5.3 The effects on human kind of acute exposure to radiation

	Dose received, Sv (rem)[a]					
	>50 (>5000)	50-10 (5000-1000)	10-6 (1000-600)	6-2 (600-200)	2-1 (200-100)	<1 (<100)
Death at	2 days	2 weeks	2 months	2 months	-	-
Cause of death	Respiratory failure; brain oedema	Loss of fluid and essential salts	Haemorrhage; infection	Haemorrhage; infection	-	-
Incidence of death	100-90%	100-90%	100-80%	80-0%	0%	0%
Principal affected organs	Central nervous system	Gastrointestinal tract	Haematopoietic tissue	Haematopoietic tissue	Haematopoietic tissue	None
Characteristic signs	Vomiting; convulsions; tremor; ataxia;[b] lethargy	Vomiting; diarrhoea; fever	Severe leukopaenia;[c] purpura;[d] haemorrhage; infection; loss of hair	Severe leukopaenia;[c] purpura;[d] haemorrhage; infection; loss of hair	Moderate leukopaenia[c]	None

[a] Equivalent to Gy(rad) for γ and β irradiation.
[b] Inability to co-ordinate muscular actions.
[c] Decrease in number of leucocytes in the blood.
[d] Bruising of the skin.
Source: Eisenbud, 1973 as modified from Glasstone, S. and Dolan, P. J. 1962: The Effects of Nuclear Weapons. 2nd edition. Washington, DC: US Department of Defense and Department of Energy

radioactive gas during the course of their work has been studied in a number of countries. Finally, patients irradiated in the course of frequently repeated diagnostic examination or during treatment of their diseases by radiotherapy can be studied. In the former group, patients receiving repeated diagnostic examinations during the treatment for pulmonary tuberculosis have sometimes received substantial radiation exposure to the chest and female patients have been found to develop breast cancer in excess of normal expectation (UNSCEAR, 1977, 1982). From these extensive and varied surveys the detrimental effects of ionizing radiations are found to be somatic (manifest in the exposed individuals themselves) and genetic or hereditary (affecting descendants). The effects may be non-stochastic (the severity of the effect varies with the dose, and for which a threshold of dose may therefore occur below which no detrimental effects are seen) or stochastic (the probability of an effect occurring, rather than its severity, is regarded as a function of dose, without a threshold). Hereditary effects are regarded as being stochastic.

It has become clear that the most important late somatic effect of low doses of radiation is the occasional induction of malignant diseases, as shown by their increased incidence in exposed populations. In general, it appears that relatively high cancer induction rates apply to the breast in females and to the thyroid, although the mortality rate from induced thyroid cancers is low – that is, about 3 per cent per 25 years. The induction rates for lung cancer and for leukaemia are somewhat lower, and those for other organs for which estimates are obtainable appear to be lower still (UNSCEAR, 1977). Typically there is a long latent period between radiation exposure and detection of a tumour, which is because of the time required for sufficient increase in the size of the tumour to make it detectable, and it may also be due in part to a form of induction period before the initially affected cell or cells start to divide and form a tumour, or before the tumour assumes 'malignant' characteristics of growth and spreading. Mean latent periods for malignant tumour detection after radiation exposure are estimated at about 25 years, although for leukaemia it is between 8–14 years.

In addition to cancer inductions, ionizing radiation may have significant effects on pre-natal development and on genetic or hereditary factors. Irradiation of the embryo or foetus may cause defects of a wide range of severity, from death while still in the uterus (based on animal experiments), impairment of growth, structural changes (or malformations) or functional deficiencies. Thus, for example, Japanese children who had been exposed in utero as a result of atom bomb detonations at Hiroshima and Nagasaki have shown a possible increased incidence of mental retardation, associated with small head-size (microcephaly) and clear evidence of reduction in body size. It may be concluded that the developing embryo and foetus show a pronounced sensitivity to the induction of malformations by radiation, particularly during the main phase of organogenesis (UNSCEAR, 1977).

Genetic effects of radiation are likely to be predominantly due to damage induced in the DNA molecular structure. When cells are exposed to ionizing radiation, the chromosomes of the cell nuclei may be damaged by the production of gene mutations (dominant and recessive), involving alterations in the elementary units of heredity which are localized within the chromosomes, or by the induction of chromosome aberrations, consisting of changes in the structure or number of the chromosomes. When such changes are induced in the germ cells, they may be transmitted to descendants of the irradiated persons. UNSCEAR (1977) estimates that in a million liveborn children in the first generation of offspring of a population exposed to 10 mGy (1 rad) at low dose rate during the generation, there would be sixty three genetic diseases. The total genetic damage expressed over all generations (or the value in each generation reached after prolonged continuous exposure) is estimated to be 185 per million per 10 mGy (per rad). However, these estimates do not take full account of the class of mutational events which lead to minor deleterious effects and which, by their large number, might impose a greater total genetic burden on the population than that from a smaller number of relatively more serious conditions.

In order to limit the frequency of stochastic effects to a low level and to prevent the occurrence of non-stochastic effects, the annual dose limit of 50 mSv (5 rem) is recommended by the International Commission on Radiological Protection (ICRP). This recommendation provides a safety margin for non-stochastic effects as it believes that 0.5 Sv (50 rem) per annum would suffice for all tissues except the lens (irradiation leads to lens opacification that would interfere with vision) for which a limit of 0.3 Sv (30 rem) in a year applies (ICRP-26, 1977; Pentreath, 1980). The estimated risk factors for radiation-induced fatal cancers and serious hereditary damage are given in table 5.4. The risk factors are mean values although some are clearly age- or sex-dependent: for example, the risk for the development of breast cancer is $5.0 \ 10^{-3} \ Sv^{-1}$ in females and zero for males. Included in this table are the 'fractional weighting factors' which quantify the fractional risk that each tissue contributes to the overall risk and allows one to make an estimate of the risks from partial body radiation. These values highlight that some tissues are more susceptible to damage than others. In addition to recommending limits of radiation doses the ICRP urges that all radiation exposures should be kept 'as low as reasonably achievable' (ALARA), economic and social factors being taken into account; and that no practice involving possible radiation release should be adopted unless its introduction produces a positive net benefit.

Table 5.4 Estimated risk factors per sievert of radiation for radiation-induced fatal cancer and serious hereditary damage

Tissue/organ	Risk factor (Sv^{-1})	Fractional weighting factor[c]
Testes, ovaries (hereditary risk)[a]	4×10^{-3}	0.25
Red bone marrow[b]	2×10^{-3}	0.12
Bone surfaces	5×10^{-4}	0.03
Lung	2×10^{-3}	0.12
Thyroid	5×10^{-4}	0.03
Breast	2.5×10^{-3}	0.15
All other tissues	5×10^{-3}	0.30
Whole body	1.65×10^{-2}	1.00

[a]Hereditary risk factor applies to the first two generations. For all generations the risk factor is 8×10^{-3} Sv^{-1}.
[b]Specified because it is the tissue mainly involved in radiation-induced leukaemia.
[c]This refers to the fractional risk that each tissue contributes to the overall risk when the body is irradiated uniformly. Application of the weighting factors allows one to make an estimate of the effective radiation dose, and hence the risk, from partial body irradiation (e.g. where radionuclides enter into the body).
Source: ICRP-26, 1977

5.3   Major and Potential Sources of Radiation

5.3.1   Natural-environmental Sources

The major source of radiation to which people are exposed is the natural environmental component (table 5.5). The contribution by human technological developments is relatively small at present but concern arises because radiation from human sources is ever increasing and because humanity has the potential, through accidental releases of radioactivity from nuclear reactors and through nuclear explosions, of creating substantial increases in local, regional and global levels of radioactivity.

Radiation exposure to natural sources is a result of both terrestrial and cosmic irradiation and varies spatially because of differences in altitude and radionuclide distribution in the terrestrial environment. Exposure to natural radiation may be enhanced by human activities such as high-altitude aircraft (and space) flights; construction of buildings using materials of high radium content (pumice stone, granite or light concrete derived from alum shale); the irradiation resulting from the phosphate industry; from using water from deep wells bored into radon-rich water; from a variety of consumer products containing radionuclides (radio-luminescent time-pieces, compasses, smoke detectors, luminous signs, antistatic devices, and colour television sets); or the irradiation due to the release of natural radionuclides from coal-fired power plants. Such

Table 5.5 Global dose commitments from various radiation sources

Source of exposure	Global effective dose-equivalent commitment[a] (days equivalent to natural radiation)
One year exposure to natural sources	365
One year of commercial air travel	0.1
Use of one year's production of phosphate fertilizers at the present production rate	0.3
One year global production of electric energy by coal-fired power plants at the present global installed capacity (1000 GW(e))	0.07
One year production of nuclear power in 1981 (about 90 GW(e)·a)	1
One year of nuclear explosions averaged over the period 1951–81	50
One year's use of radiation in medical diagnosis	70

[a]The global dose commitment for each of these radiation sources is expressed as the duration of exposure of the world population to natural radiation which would cause the same dose commitment. The occupational contribution is included.
Source: Adapted from UNSCEAR, 1982

plants produce significant radionuclides (uranium and daughters, thorium and daughters) from coal of various origins and in slag and fly-ash, exposing people to external radiation (from discharged material deposited on the ground) and to internal radiation (from the material inhaled). The radioactivity comes from radioactive elements in the coal. As the coal burns, these radioisotopes concentrate because the organic matter in the coal perishes. It is even suggested that in the United States the routine exposure of the population to radionuclide release from coal burning in that country could be significant and possibly comparable with the effects from nuclear power (Wilson et al., 1980).

5.3.2 Sources from Human Technological Developments

5.3.2.1 Radiological sources The major sources of ionizing radiation arising from human technological developments include medical uses of radiation, nuclear explosions, and power generation from nuclear fission. The highest human-derived contribution to global collective dose is caused by the medical uses of radiation, and in particular by diagnostic X-ray procedures. At present 5×10^5 man-gray is produced by a small number of the world's countries with developed radiological facilities, while 2×10^4 man-gray come from the majority of the world's population who live without the benefit of frequent radiological examinations (Coggle, 1983). Diagnostic radiology is growing at a rate of between 5 and 15 per cent per year in many technically-developed countries which, together with the rapid growth rate expected in developing countries, will lead to substantial increases in the global dose commitment in the future.

5.3.2.2 *Nuclear-weapons testing*   Atmospheric testing of nuclear weapons produces global contamination and the most important radio-isotopes of fallout are those which give an external gamma-ray dose and those which become internally deposited in the body, either by direct absorption or via a food chain. Following a nuclear weapons test the residence time of the radionuclides released depends upon the height to which the radioactive cloud rises (which is related to explosion yield), its latitude of detonation (the height of the tropopause being lower in the polar regions than near the equator), the size of the radioactive aerosols, and whether the radionuclides are in gaseous or particle form. Large particles will fall out within a few hundred kilometres of the detonation site (the local fallout) while smaller particles and gaseous radionuclides injected into the troposphere may be transported around the Earth in the same hemisphere and between hemispheres by for example, the East African jet stream (Findlater, 1974), to be deposited hundreds or even many thousands of kilometres away (the tropospheric fallout). Particles and gases which are injected into the stable stratosphere may take months or years (for example, the residence time of strontium-90 is typically of the order of a year) before they become stratospheric fallout. Ironically, the longer radionuclides remain in the stratosphere, the weaker will be the radioactivity when they return to the Earth's surface.

Atmospheric tests by the United States, the Soviet Union, and the United Kingdom ceased with the signing of the Partial Test-Ban Treaty of 1963 although other countries such as France and China continue to undertake such tests. The total global dose commitment from all nuclear explosions carried out before 1976 ranges from about 1000 μGy in the gonads to about 2000 μGy in bone-lining cells (UNSCEAR, 1977). In the northern temperate zone the values are about 50 per cent higher than these estimates, and in the southern temperate zone, about 50 per cent lower. External exposures contributed by caesium-137 and short-lived gamma-emitting radionuclides account for about 700 μGy of the global dose commitment for all tissues. Internal exposures are dominated by contributions from the long-lived radionuclides caesium-137 and strontium-90 (in the skeleton). Their half-lives of about 30 years will determine the length of time over which the doses will be delivered. The more short-lived

Table 5.6 Nuclear power stations in operation and under construction at the end of 1985

Country	Reactors in operation		Reactors under construction		Nuclear electricity supplied in 1985		Total operating experience to end 1985	
	No. of units	Total Mw(e)	No. of units	Total Mw(e)	TW(e).h	% of total	Years	Months
Argentina	2	935	1	692	5.2	11.3*	14	7
Belgium	8	5486			32.4	59.8	64	1
Brazil	1	626	1	1245	3.2	1.7*	3	9
Bulgaria	4	1632	2	1906	13.1	31.6	30	6
Canada	16	9776	6	4789	57.1	12.7	151	7
China			1	300				
Cuba			2	816				
Czechoslovakia	5	1980	11	6284	10.9	14.6	22	4
Finland	4	2310			18.0	38.2	27	4
France	43	37533	20	25017	213.1	64.8	338	5
Germany (GDR)	5	1694	6	3432	12.2*	12.0*	57	5
Germany (FRG)	19	16413	6	6585	119.8	31.2	215	4
Hungary	2	825	2	820	6.1	23.6	4	5
India	6	1140	4	880	4.0	2.2*	54	8
Iran			2	2400				
Italy	3	1273	3	1999	6.7	3.8	69	10

	Operating No.	Operating MW	Construction No.	Construction MW	TWh	%	Years	Months
Japan	33	23665	11	9773	152.0	22.7	286	10
Korea RP	4	2720	5	4692	13.9	22.1*	15	5
Mexico			2	1308				
Netherlands	2	508			3.7	6.1	29	9
Pakistan	1	125			0.2	0.9	14	3
Philippines			1	620				
Poland			2	880				
Romania			3	1980				
S. Africa	2	1840			5.3	4.2	2	3
Spain	8	5577	2	1920	26.8	24.0	56	10
Sweden	12	9455			55.9	42.3	99	2
Switzerland	5	2882			21.3	39.8	53	10
Taiwan (China)	6	4918			27.3*	53.1*	26	1
UK	38	10120	4	2530	53.8	19.3	695	10
USA	93	77804	26	29258	383.7	15.5	954	11
USSR	51	27756	34	31816	152.0	10.3*	531	7
Yugoslavia	1	632			3.9	5.1*	4	3
TOTAL	374	249625	157	141942	1401.6		3825	3

*IAEA estimates.
Source: International Atomic Energy Agency, 1986

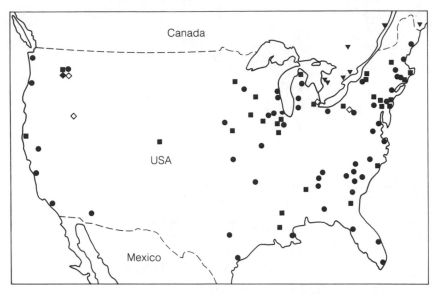

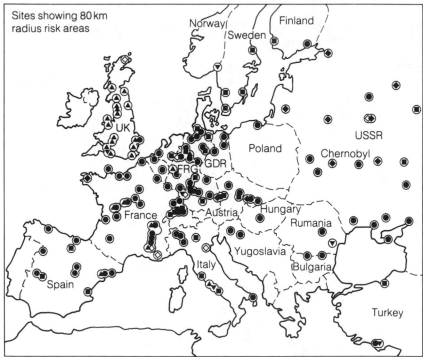

Figure 5.4 Nuclear power stations in North America and Europe
Source: Hawkes et al., 1986

ruthenium-106 and cerium-144 are significant contributors to the exposure of the lung. The irradiation to which the world population was committed by nuclear tests in the 1950s and early 1960s is already completed for all except caesium-137 (external and internal), hydrogen-3, carbon-14, strontium-90, and plutonium-238. Caesium-137 makes major contributions to all body tissues which are essentially independent of age, whereas the further substantial contribution from strontium-90 in bone marrow and bone-lining cells is strongly dependent on age. In comparison, the contributions of the other radionuclides to current annual doses are very small. Obviously, the irradiation from past atmospheric weapons tests is minute compared with that which would follow a large-scale nuclear war (Ehrlich et al., 1983; Turco et al., 1983).

5.3.2.3 Nuclear power generation   The generation of power from nuclear fission, generating intense radioactivity, has been developed and used commercially for three decades to meet part of the world demand for electrical energy. Most of the world's nuclear reactors are termed thermal reactors and are fuelled by natural or enriched uranium. The Pressurized Water Reactor (PWR) predominates in most national nuclear power industries with some notable exceptions, including the UK (some Advanced Gas-Cooled Reactors or AGRs but mostly Gas-Cooled Reactors or GCRs), Canada (the CANDU type of reactor which uses heavy water as the coolant), and the USSR (half are of the Light Water Graphite Reactor or LWGR type). More recently fast-breeder reactors fuelled with a mixture of plutonium and uranium have been developed by some countries. By the end of 1985, 374 nuclear power reactors were in operation in twenty six countries generating more than 18 per cent of the world's electricity (figure 5.4; table 5.6). An additional 157 reactors are at present under construction.

Key to reactor types	
▲ GCR Gas cooled reactor Magnox Magnox type gas cooled reactor AGR Advanced gas cooled reactor HTGR High temperature gas cooled reactor	■ BWR Boiling water reactor ● PWR Pressurized water reactor
◇ FBR Fast breeder reactor LWGR Light water breeder reactor	▼ PHW Candu Pressurized heavy water Candu BLW Candu Boiling light water Candu BHWR Boiling heavy water reactor SGHWR Steam generating heavy water reactor
◆ GCHWR Gas cooled heavy water reactor LWGR Light water cooled graphite reactor	PHWR Pressure vessel heavy water reactor LWCHWR Light water cooled heavy water reactor

Key to Figure 5.4

In 1982, the International Atomic Energy Agency estimated that installed nuclear capacity would amount to between 750 and 1100 GW(e) by the end of the century. Such an estimate is significantly lower than El-Hinnawi's calculation in 1980, when nuclear energy capacity by the turn of the century was projected to be between 900 and 2000 GW(e). Projections for the future are subject to great uncertainty because of the factors involved in energy forecasts and because of uncertainty in the willingness of the public to accept the potential risks associated with increased use of nuclear energy. Recent reactor accidents have caused a marked slow-down in the installation of nuclear power stations. Table 5.6 reveals that some countries, with small resources in conventional energy sources, including Belgium, Bulgaria, Finland, France, Sweden, Switzerland and Taiwan rely heavily on nuclear energy. The share of nuclear power for electricity generation in these countries exceeded 30 per cent in 1985 but may reach 70–80 per cent by the year 2000 unless, as in Sweden, nuclear energy production loses favour.

Nuclear power production involves a series of steps comprising the processes of mining and milling of uranium, conversion to fuel material (in most cases including enrichment in the isotope uranium-235 from about 0.7 per cent to 2–4 per cent), fabrication of fuel elements, utilization of the fuel in nuclear reactors, storage of the spent fuel, reprocessing of this fuel with a view to recycling, transportation of materials between various installations and disposal of radioactive waste (figure 5.5). Almost all of the radioactive material associated with the nuclear industry is at present in the reactors and in spent fuel or in well-contained fractions separated from the fuel during the reprocessing operations. However, at most steps of the operations, releases occur either through the gaseous or liquid effluent streams (UNSCEAR, 1977, 1982). Most of the radionuclides released are of local or regional concern, because their half-lives are short compared with the time required for dispersion to greater distances.[1] However, some radionuclides having longer half-lives or being more rapidly dispersed can become globally distributed. Although at present, nuclear power generation contributes only 0.1 per cent of the total global radiation dose commitment, much public concern has been expressed on the potential local and regional effects of accidental releases of radioactive material.

5.3.2.4   Nuclear accidents   Although accidents involving the release of radioactive material can occur at any stage of the nuclear fuel cycle, most

[1]Although the concern about ionizing radiation centres on health effects, Boeck (1976) has suggested that the routine releases of krypton-85 from nuclear power stations may lead to increased ionization of the atmosphere which may weaken cloud formation processes and cause a decrease in global precipitation.

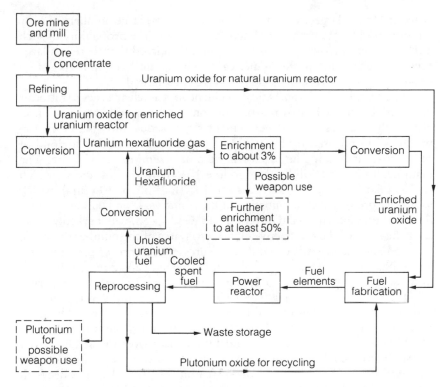

Figure 5.5   Steps involved in nuclear power production
Source: Coggle, 1983

attention has been focused on reactor accidents. This is not because reactor accidents are more likely, but because the potential consequences for the general public of some reactor accidents are much greater than from accidents at other stages in the fuel cycle.

Several studies have been made in which the probability of reactor accidents of various degrees of severity were estimated, and in which the environmental impacts of the radiation releases associated with these hypothetical accidents were calculated. The type of accident considered involves the reactor safety features that control the cooling and running of the reactor. If they fail, the temperature of the reactor core rises and the fuel melts and may breach the main containment above the reactor. Such a 'meltdown' results in the release of large quantities of radioactive material. It needs to be stressed that in a thermal reactor the fuel is near to its most active state and there is no way in which the nuclear assembly can go supercritical and explode like a nuclear weapon (Coggle, 1983). In contrast, the plutonium fuel used in fast-breeder reactors could possibly go supercritical during a core callapse meltdown. In general, studies have suggested that a large accidental release could cause large numbers of early fatalities and large numbers of latent cancers, but the probability of such a release is low. However, it is now generally agreed that there is a large range of uncertainty in the numerical results quantifying the risks of an accident as recent reactor accidents have highlighted.

One example of a hypothetical accident at a nuclear power station is provided by the UK Royal Commission on Environmental Pollution (1976). It assumes that 10 per cent of the gaseous and volatile fission products are released from a 100 MW(e) nuclear reactor as a cloud of radioactive material. The reactor is sited at a semi-urban UK site. The main health hazard will be from iodine-131 (half-life of 8 days), which will irradiate the thyroid, and caesium-137 (half-life of 30 years), which will cause prolonged contamination of the countryside and buildings. Since the weather, especially wind direction and wind speed, will markedly affect the behaviour of the radioactive cloud, estimates of the number of people exposed and the hazards can only be expressed in terms of probabilities. The inhalation of iodine-131 could cause thyroid cancer in people as far away as 24 km and there is a 20 per cent probability that over a period of 10–20 years, between 1000 and 10,000 people could develop thyroid cancer (less than 10 per cent of these would be fatal cancers). The most probable outcome is 100–150 deaths from thyroid cancer plus a further 10–200 deaths from leukaemia and lung cancer over the same period. These figures could be ten times higher or lower depending on the circumstances. The iodine-131 deposited on the ground would decay in a few weeks. However, the longer-lived caesium-137 would be present in the soil and buildings for many decades. The radiation levels could necessitate evacuation for weeks, months or longer, even for people up to 50 km from the reactor site. Similar theoretical studies have been made in other countries such as that by Professor Norman Rasmussen in the United States, who concluded that the worst possible reactor accident would cause 3300 deaths, 45,000 illnesses, and 1500 fatal cancers which would appear decades later (Hawkes et al., 1986).

Serious accidents have taken place in the nuclear fuel industry, most notably with the fire at Windscale (renamed Sellafield in 1981) in northwest England in 1957, the overheating and core and containment damage at the PWR at Three Mile Island, USA, in 1979, and at Chernobyl in the Soviet Union in April 1986. At Windscale a sudden temperature rise and ignition of the uranium fuel cartridges within an air-cooled graphite moderated reactor (a type designed solely for the production of military plutonium) led to the release of ionizing radiation from the reactor stack. The most important radionuclide released as far as health consequences to the population were concerned was iodine-131 (a gamma emitter that can cause thyroid cancer). The radioactive cloud was shown to have drifted over England, Wales and parts of continental Europe (ApSimon et al.,

1985; Chamberlain, 1959). The group of the population which was most at risk was young children drinking locally produced milk. Over an area of 520 km² restrictions were imposed on the distribution of milk which contained more than 3700 Bq 1^{-1} (0.1 μCi/1) of iodine-131. The maximum measured individual thyroid dose was 160 mSv to a child in the Windscale area. The milk restriction was lifted 44 days after the accident when the radioactivity levels fell to an acceptable level (iodine-131 has a half-life at 8.1 days). Other items of diet, including eggs, vegetables, meat and water were examined in the most contaminated regions, but it was decided that there was no appreciable health hazard from these foodstuffs and hence that there was no need to implement bans on them. The collective effective dose equivalent commitment from the release is estimated to have been 1.2×10^3 man Sv.

Rather belatedly, Crick and Linsley (1982) estimated that a total of 20 'health effects' (cancer deaths plus hereditary effects in the first two generations) would have resulted from the accident. Thirteen of the health effects would be due to thyroid cancer fatalities, assuming a 5 per cent fatality rate among a total number of expected thyroid cancers of 250. However, Urquhart (1983) argues that this estimate ignores the contribution of polonium-210 (used by the USA and UK as an irreplaceable initiator for starting the chain reaction at the heart of their first atomic bombs) to the health effects and that the radiation-induced cancer deaths and hereditary effects would have been considerably higher. In this connection, he claims that an increase in the number of myelomas in the area during the 1970s would not have been expected as a result of iodine-131 contamination, but could have arisen from ingestion of polonium-210, an alpha emitter.

The next known serious nuclear-reactor accident occurred at Three Mile Island, Harrisburg, Pennsylvania on 28 March 1979. One of the two pressurized water reactors initially suffered the loss of normal feedwater supply which led to a turbine trip and later to a reactor trip which subsequently resulted in significant damage to portions of the reactor core. It is believed that the sequence of events that led to core damage involved equipment malfunctions, design-related problems, and human errors, all of which contributed in varying degrees to the accident (US NUREG-0600, 1979). The principal radioactive materials released into the environment were xenon-133 (half-life of 5.3 days), xenon-135 (half-life of 9.2 hours), and traces of radioactive iodine, primarily iodine-131. Some of the radioactive krypton isotopes with half-lives of a few hours may also have been released (US NUREG-0558, 1979). Radionuclides in particulate form such as strontium-90, uranium isotopes, and plutonium were not detected in the environment and would either have been retained in the fuel or, if released from the fuel, would have remained in the coolant water.

The collective dose to the total population (about 2 million) within an 80-kilometre radius of the Three Mile Island plant has been estimated to be 3300 person-rem (for the period from 28 March to 7 April) with an average dose to individuals of 1.5 mrem. An early estimate of the number of excess health effects (cancer and genetic ill-health) due to the irradiation of this population is approximately two (US NUREG-0558, 1979). Despite the apparently small health effects, this accident had far-reaching consequences: the credibility of the safety record of the nuclear-power industry fell markedly. People realized that however sophisticated and advanced a technology may be there is no fool-proof technology and no safeguard against human error – and nor, it should be added, against terrorist or military intervention. NUREG has subsequently calculated that there is a 45 per cent chance of an accident similar to Three Mile Island happening in the next 20 years.

If the public's confidence in the safety record of the nuclear-power industry faltered following the Three Mile Island accident, the accident at Chernobyl in the Soviet Union caused nuclear reactor safety to be completely re-examined. After only 4000 reactor years worldwide, Chernobyl approached the worst conceivable nuclear reactor accident, which was previously estimated to be of a probability somewhere near 1 in 10,000 or even 1 in a million per reactor year. The Chernobyl accident released 50–100 million curies of radioactivity, according to the official Soviet report, or as much as 287 million curies of radio-isotopes with half-lives greater than one day according to Hohenemser et al. (1986). This makes it 100 times worse than the Windscale accident and 1000 times worse than Three Mile Island accident. The accident at the graphite moderated, light water-cooled pressure tube reactor at Chernobyl occurred on 26 April 1986 when an explosion produced an uncontrollable fire which lasted several days and led to vast quantities of radionuclides being lifted high into the atmosphere. The plume of radionuclides was swept across Europe during the following 7–10 days, exposing up to 400 million people in 15 nations to high levels of ionizing radiation. In areas where heavy rainfall scavenged the radionuclides from the atmosphere, radiation levels peaked up to several hundred times higher than background levels.

The accident resulted in over 30 Soviet citizens dying from radiation sickness (the majority of fatalities–mostly reactor workers or firemen–had received whole-body doses ranging from 600 rems to 1600 rems). Another 1000 people were treated for radiation sickness, some of whom needed bone-marrow transplants to restore their white blood-cell count. The number of casualties near the reactor was less than might be expected because the radiation plume was made especially buoyant by the intense fire and so, like the effect of a tall industrial stack, local deposition of radionuclides was reduced whereas the long-range transport of radio-nuclides was enhanced. Within a 30 km radius of Chernobyl, 135,000 people were evacuated. Evacuation was not prompt because the Soviet authorities were initially unwilling to publicize the seriousness of the accident and because few Soviet citizens had access to private transport. These people now face annual medical inspection for the rest of their lives to assess what the radiation may have bequeathed to them in terms of a legacy of cancer or birth defects.

Estimates of the number of expected cancers resulting from exposure to the radionuclides vary quite markedly, according to the assumptions made in the model employed (Wilson, 1986). Estimates range from several thousands to tens of thousands of thyroid cancers arising from the exposure to iodine-131 and a similar number of cancers elsewhere in the body from caesium-137. Only a few per cent of the thyroid tumours would be fatal, whereas perhaps half of the cancers from caesium-137 would be fatal. Hawkes et al. (1986) estimate the likely number of deaths in the Soviet Union at between 5000 and 10,000. The official Soviet estimate of the projected cancer deaths in western Russia after Chernobyl is put at 4000–5000 due to exposure to iodine-131 and as many as 40,000 from exposure to caesium-137. Some researchers, pointing out that no radiation dose is so small that the body can perfectly repair all resulting damage to DNA and chromosomes, claim that as many as a million people in the Northern Hemisphere may develop cancer as a result of Chernobyl and that half of these cancers would be fatal.

There are many problems involved in calculating the number of cancers and deaths resulting from the Chernobyl accident. The ground-level deposition of radionuclides is not known accurately and the existence of local hot spots of radiation, where heavy rainfall scavenged the radionuclides from the atmosphere, may not be readily incorporated into some models. Some hot spots can be very localized: for example, in Konstanz, where it rained on 30 April, ground-level radiation was 15 times greater than nearby Stuttgart, which received no rain. In many places in Europe, fallout from Chernobyl exceeded the peak of the weapons-testing fallout in 1963 (figure 5.6; Thomas and Martin, 1986). Other problems associated with estimating the health effects of the Chernobyl radiation release include the difficulty of taking into account all of the radionuclides involved and all of the protective measures adopted by governments and individuals to reduce the radiation dose people received. Protective measures included staying indoors during the passage of the radiation cloud, the removal of the outer leaves and washing of vegetables, the washing and peeling of fresh fruit, the ban by Western European countries on the importation of fresh food from areas of Eastern Europe within 1000 km of Chernobyl, the restriction in consumption of fresh milk and free-range eggs, the removal of cattle and sheep from open pasture or a ban on their sale, the taking of prophylactic potassium iodide tablets to saturate the thyroid with iodine so as to minimize the amount of radioactive

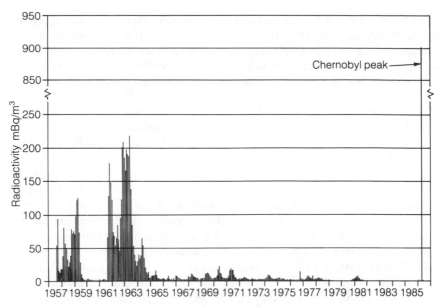

Figure 5.6   Radioactivity of dust samples collected at Uccle, Belgium, for the period 1957-86
Source: Supplied by the Royal Meteorological Institute, Brussels, Belgium

iodine from being taken in (this measure was adopted most extensively in Poland), the advise not to drink fresh rainwater, and the warning to the public about possible contamination during outdoor activity.

As with most other air-pollution disasters, the Chernobyl accident highlighted many inadequacies in the way in which the authorities reacted to the accident. The Soviet government failed to provide a warning of the accident to European governments and when information was offered it was incomplete. Many European governments, such as France and the United Kingdom, failed either to provide adequate answers to the public's urgent questions or to introduce precautionary measures quickly enough. Radiation monitoring networks were inadequate in many European countries for the purpose of rapidly pinpointing radiation hot spots, though in hindsight a rainfall map of Europe for the period provided a useful approximation.

Such criticisms do not merely require that governments react more effectively next time, but that there is a need to ensure that a Chernobyl-type accident does not happen again. Some would argue that while nuclear-power reactors exist, then a repetition is inevitable. However sophisticated the safety designs incorporated into nuclear power stations, they cannot overcome the human fallibility which can occur. In both the Three Mile Island and Chernobyl accidents, the operators misread the reactor's condition and shut off the emergency systems at the wrong moment.

Similar human error surfaces in other air-pollution disasters such as at Seveso and Bhopal. What the public and governments have to assess is the need for nuclear energy versus the risk of another Chernobyl-type accident. To the concern for another serious reactor accident, one has to add the concern for accidents during other stages of the nuclear fuel cycle, the long-term risks associated with the safe disposal of ever-increasing radioactive waste, and the decommissioning of nuclear-power plants at the end of their 30-year operating life (about 100 such stations by the end of the century).

Following the Three Mile Island accident and especially after the Chernobyl accident, various responses have been made by national governments and the International Atomic Energy Agency. The tightening-up of safety standards at nuclear-power plants inevitably followed the reactor accidents. The need for greater international collaboration so as to be able to respond to reactor accidents which have no respect for international borders was urged. The International Atomic Energy Agency (IAEA), based in Vienna and with 113 member countries, was given a stronger role. The Agency organized two international agreements, prepared since 1982, but signed in September 1986 as a result of the Chernobyl accident. The first convention is concerned with the rapid notification of nuclear accidents (even at a military installation). The need for neighbouring countries to receive such quick information is obvious, given the tendency for countries to site their nuclear reactors near their borders. The second convention covers the offering of mutual assistance in the event of an accident.

While such action is welcome, for some governments and for many people, no amount of safety measures or reassurances regarding the ability to cope with a reactor accident is satisfactory. Public anxiety following Three Mile Island and Chernobyl was especially evident in the weeks following these accidents as demonstrations became frequent throughout Europe. Such public pressure, sustained by the media, became translated into political pressure in many countries. This is why Sweden has decided to phase out its twelve nuclear-power stations by the year 2010 (even though they currently provide more than 40 per cent of the country's electricity), Austria has decided to dismantle its only nuclear-power station, the Philippines has halted construction of its first reactor, and countries such as Australia and Denmark have declined to build any nuclear-power plants. Several other countries have either scaled down their original expansion plan for nuclear power, delayed the continuation of their nuclear-power expansion or have begun to look seriously at the phasing out of existing nuclear-power stations. However, for countries such as France, Japan and Taiwan and a number of developing countries which lack readily available alternative energy sources, the development of nuclear energy still remains a priority. In Poland, where coal reserves are large, the government views the development of nuclear energy as necessary to satisfy the nation's energy needs. The Soviet Union's commitment to nuclear power remains undiminished. It seems that accidents such as Three Mile Island and Chernobyl have initiated a move away from nuclear power but that some countries still remain fully committed to its development.

6 Global Pollutants: Carbon Dioxide and Ozone Depletion

6.1 Compounds Depleting Stratospheric Ozone

The atmospheric 'ozone layer' is a layer of relatively high concentration of ozone (O_3) located in the stratosphere. Maximum ozone concentration occurs at a height of 16–18 km in polar latitudes and at about 25 km over the equator. This ozone layer is important to society and the environment for two reasons. Firstly, the ozone absorbs solar ultraviolet radiation in the UV-B or 280–320 nm wavelengths, thus warming the stratosphere and producing a steep temperature inversion between 15 and 50 km. As temperature decreases with height in the troposphere (reaching -40 to $-80°C$ at the tropopause), the effect of a warm stratosphere is to act as a ceiling or lid to vertical motion in the troposphere – that is, convective processes which produce clouds and precipitation are contained within the troposphere. Any weakening of the stratospheric inversion would affect convective processes and atmospheric circulation in general, thereby affecting weather and climate. A reduction of ozone would also cause more ultraviolet and visible radiation to reach the ground, so leading to a warming of the lower atmosphere and the Earth's surface. At the same time, reduced absorption of the ultraviolet radiation in the stratosphere would reduce heating there by perhaps $10°C$. This situation would in turn tend to promote surface cooling, as less thermal radiation would be emitted from the stratosphere to the ground. These opposing effects, one of warming and the other of cooling, would be complicated by the changed distribution of ozone in the stratosphere such that the net effect is difficult to predict. The UK Department of Environment (1979a) suggests the effect of ozone depletion on surface temperature would be barely detectable while Reck (1976) suggests that the surface temperature would change by between $-0.6°C$ and $+0.8°C$ – the sign depending upon the underlying surface albedo and the presence or absence of clouds and particulate pollution layers (warming when aerosol hazes are present, cooling when not present). Increasingly, research tends to favour a troposphere warming due to ozone depletion (US National Research Council, 1982a).

The second reason why the ozone layer is important is because through absorption, it controls the amount of ultraviolet (UV-B) or erythemal solar radiation reaching the ground. UV-B radiation has both beneficial and harmful effects on plants, animals and people. On the beneficial side, the whole population gains from the action of UV-B radiation in converting steroids in the skin to vitamin D. On the harmful side, it is known to affect the growth, composition, and function – including photosynthesis – of a wide variety of plant species, including many important crops such as soya beans. Experimental studies with UV-B levels enhanced from zero to 50 per cent above natural levels have shown adverse effects on fishes' eggs, larvae and juveniles, on shrimps, crabs, zooplankton, and other aquatic organisms, as well as on phytoplankton and other aquatic plants which are essential to the aquatic food-web. In animal studies, UV-B radiation has been shown not only to be carcinogenic, but also to alter the response of the immunological system; this results in impeded recognition of a UV-B tumour as a foreign body.

In people, UV-B radiation is known to cause ageing of the skin, and because of the high correlation between sunshine and skin cancer, it is accepted that UV-B exposure of skin is directly linked to skin cancer (Eaglemann, 1981). The most serious form of skin cancer is malignant melanoma, which has a 30 per cent fatality rate (Panofsky, 1978). Malignant melanoma of the skin is caused by cancerous proliferation of melanocytes. Melanocytes are those cells that produce the brown pigment, melanin, which is responsible for the principal differences in the colours of the human race. Also, melanin is the pigment that results in protective tanning after exposure to UV-radiation (US National Research Council, 1984). Calculations on the effect of percentage changes in total ozone column on percentage changes in UV-B radiation, using the absorption and scattering properties of ozone and other atmospheric constituents, have indicated an average amplification or magnification factor between 1.5 and 3.0 – depending upon latitude and season. In other words, a 5 per cent reduction in ozone would result in, say, an increase of between 7.5 and 15 per cent in UV-B radiation received at the ground. Such an effect has been estimated to cause an increase in the incidence of basal cell and squamous cell (nonmelanoma) skin cancer by tens, or even, hundreds of thousands of cases a year in the United States. Basal cell and squamous cell carcinomas appear predominantly on exposed areas of the head and neck, while in Caucasians, malignant melanomas frequently appear on the back in both men and women and also in the legs in women. As skin cancer frequencies increase latitudinally towards the equator, correlating with increasing UV-B exposure, this projected increased incidence in skin cancer is equivalent in terms of exposure to the population living several degrees of latitude closer to the equator.

Given the significance of the stratospheric ozone layer, concern regarding the effects of pollution of the stratosphere has been expressed in recent years. Ozone is produced at stratospheric levels by photodissociation of molecular oxygen by radiation wavelengths less than 242 nm and removed by the classical reaction of $0 + O_3$ at levels below the stratosphere. However, ozone also participates in complex chemical reactions involving trace-substances in the stratosphere. Human activities are increasing the concentration of substances such as oxides of nitrogen, oxides of hydrogen, chlorine and bromine, to the extent that, through catalytic cyles, these substances may be removing ozone at a faster rate than it is being produced. Stratospheric chemistry is very complex, involving more than 150 chemical reactions between 50 chemical species, so it is by no means certain that all potential threats to the stratospheric ozone layer are yet known. Possible causes of ozone depletion identified to date include various combustion products emitted from supersonic aircraft; nitrous oxide released from nitrogen-based chemical fertilizers; oxides of nitrogen produced by nuclear-weapon testing; and chlorofluorocarbons used in aerosol sprays, refrigeration systems and industrial processes.

6.1.1   Supersonic-aircraft Emissions

Concern for the ozone layer was first voiced when large fleets of the Anglo-French Concorde, The Soviet TU-144, and the American Boeing 2707 supersonic stratospheric-flying aircraft (SSTs) were planned (Sidebottom, 1979). It was argued that the various combustion products such as soot, hydrocarbons, oxides of nitrogen, sulphate particles, and water vapour introduced directly into the stratosphere would, because of their long residence-time (months, years, decades), participate in photochemical reactions leading to a reduction in ozone concentration. Early emphasis was given to water vapour (Harrison, 1970; Newell, 1970) but soon afterwards, oxides of nitrogen became the primary focus (Crutzen, 1972; Johnston, 1971). It was calculated from relatively simple one-dimensional models that 500 Concorde-like SSTs would cause approximately a 3 per cent ozone-reduction, whereas a similar number of the planned Boeing SSTs would reduce the ozone by 15 per cent (US National Research Council, 1975a). This significant difference arose because whereas the Concorde and the TU-144 would cruise at an altitude of 17 km, the Boeing 2707 would cruise at 20 km, and in addition, the Concorde and TU-144 would consume only about one-third as much fuel as the larger, faster Boeing. Injected at 17 km, a given mass of oxides of nitrogen was estimated to have only about half as great an effect on ozone as when injected at 20 km (figure 6.1). As it turned out, the Boeing 2707 was cancelled and large fleets of SSTs have not yet materialized. Estimates of the effect of the few Concordes currently in operation are that they have a negligible impact on the ozone layer.

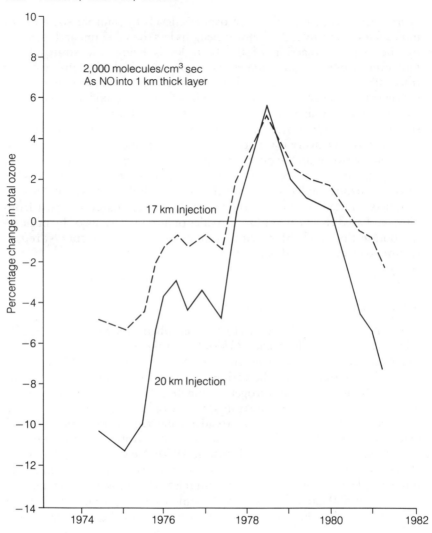

Figure 6.1   Changes in published predictions of stratospheric-ozone depletion by oxides of nitrogen by aircraft
Source: Clark et al., 1982

During the 1970s, knowledge of stratospheric chemistry improved considerably and this led to some radical rethinking of the effect of oxides of nitrogen on the ozone layer, as illustrated by the World Health Organization statements concerning the potential effect of aircraft emissions. In 1975 it was believed that a large fleet of supersonic aircraft flying at altitudes up to 25 km would have a 'noticeable effect' on the ozone layer (World Health Organization, 1976b). In 1978, this statement was revised to an 'insignificant effect' (Panofsky, 1978). By 1982, stratospheric-flying aircraft – whether supersonic or subsonic – would be expected to have a 'negative effect' on ozone – that is, causing a significant depletion of the ozone layer (Hulm, 1982). An additional finding was that oxides of nitrogen emitted by aircraft flying in the upper troposphere would shift the balance between ozone destruction and ozone production by water vapour radicals, leading to ozone production in that level of the atmosphere. Derwent (1982) suggests that current upper tropospheric aircraft operations may have already led to such an increase in the total ozone column by up to several per cent.

6.1.2   Oxides of Nitrogen from Agricultural Activities

During the mid-1970s, it was suggested that the increased use of nitrogen-based agricultural fertilizers and/or nitrogen-fixing vegetation might affect the nitrogen cycle and result in an increase in the amounts of oxides of nitrogen released from the surface into the atmosphere. This would then lead to an increase of oxides of nitrogen in the stratosphere which, through chemical reactions, would deplete ozone. However, in 1982 the World Health Organization stated that there was little likelihood of a significant change in the ozone layer in this century as a result of changing agricultural practices, but that the topic requires further study because of possible long-term effects (Hulm, 1982). During the first quarter of the next century, McElroy et al. (1976) suggest that reductions in ozone of the order of 20 per cent might result from current and future use of fertilizers. In contrast, both Crutzen (1976) and Lie et al. (1976) estimate depletions of only between 1 and 4 per cent for this time-scale.

6.1.3   Oxides of Nitrogen from Nuclear-Weapon Tests

Any process which heats the air above approximately 2300°K will produce significant quantities of nitric oxide (NO). Nuclear explosions produce shock waves which can inject oxides of nitrogen into the stratosphere. Indeed, the presence of oxides of nitrogen is indicated by the orange colour of the nuclear mushroom cloud at high altitudes, which is a consequence of some of the nitric oxide having been converted to nitrogen dioxide. The larger the explosion (threshold value of approximately 0.5 million tonnes-equivalent of TNT) and the higher the explosion above the ground, then the more that oxides of nitrogen are injected into the stratosphere. Some researchers have studied the effects of nuclear-weapon testing, which reached a peak in 1961–2, on stratospheric-ozone concentration and have arrived at contrasting conclusions. Goldsmith et al. (1973) calculated that the 1.3–1.7 million tonnes of oxides of nitrogen injected into the stratosphere as a result of 340 million tonnes of nuclear explosions in 1961 and 1962 were equivalent to the effects of between 600 and 1000 fully operational Concordes, and that there was no significant effect on the stratospheric ozone layer during or after the tests. On the other hand, Reinsel (1981) studied ozone data from a network of Dobson stations over the period 1958-79 and, in agreement with several other research groups, cautiously suggested a total ozone decrease of between 2.0 and 4.5 per cent due to nuclear-testing effects in the early 1960s. It is possible that not only did the increased stratospheric concentration of oxides of nitrogen deplete ozone in the early 1960s, but that the oxides of nitrogen also absorbed incoming radiation by as much as 6-8 per cent in some months (Kondratyev and Nikolsky, 1979). Claims made in the early 1960s– scornfully dismissed at that time – that nuclear-weapon testing was responsible for the anomalous and extreme weather and climate of that period may yet be proved to be valid.

Whether or not the testing of nuclear weapons in the early 1960s caused a significant decrease in stratospheric ozone may remain controversial but it is generally agreed by researchers that a future large-scale nuclear exchange with detonations totalling 5000-10,000 million tonnes would lead to substantial depletion of stratospheric ozone. For a 10,000 million tonne nuclear-war scenario, it is estimated that the ozone column would be reduced by between 30 and 70 per cent in the Northern Hemisphere and by up to 40 per cent in the Southern Hemisphere (Crutzen and Birks, 1982; US National Research Council, 1975b; Turco et al., 1983; Whitten et al., 1975). Such substantial stratospheric ozone depletion would last for several years and the initial survivors of a nuclear war would have to face a many-fold increase in biologically active UV-B radiation. The increased UV-B would not only suppress the immune system in people and other mammals, thereby increasing the likelihood of disease, but it would also increase the incidence of skin cancer, impair vision systems in mammals, and disrupt ocean and terrestrial ecosystems (Ehrlich et al., 1983; Elsom, 1984c, 1985).

6.1.4 Chlorofluorocarbons

The increasing domestic and industrial use of a number of stable chlorine-containing compounds has given rise to claims that these compounds may diffuse upwards into the stratosphere where they are dissociated by solar radiation to yield atoms of chlorine which act to destroy ozone through a complex chain of chemical reactions (Crutzen, 1974; Molina and Rowland, 1974). Of these compounds, special attention has been given to chlorofluorocarbons (CFCs), especially $CFCL_3$ (Freon 11) used mainly as a propellant in aerosol sprays, and CF_2CL_2 (Freon 12) used extensively as a cooling agent in refrigerators and air conditioners. CFCs are only very slowly removed from the stratosphere and with residence times in the range of 65–90 years (Brice et al., 1982), virtually all of the CFCs ever produced should still be present in the stratosphere. Because of the slow diffusion rates into and through the stratosphere, even if all emissions of CFCs into the atmosphere ceased, stratospheric concentration of CFCs would continue to increase for several years.

Estimates of the reduction in ozone concentration which may be caused by CFCs have varied from 3 to 20 per cent (figure 6.2), with the predicted rate of depletion being successively reduced since 1979 (Hulm, 1982; US National Research Council, 1982a, 1984). This reflects our increased knowledge of stratospheric chemistry such as the recognition that chlorine atoms are removed by oxides of nitrogen at a faster rate than was once believed. Whether CFCs have already significantly depleted ozone is difficult to confirm. Natural perturbing effects such as lightning, volcanic eruptions, cosmic rays, solar photon events, global circulation changes, and the variability of solar radiation, all affect stratospheric-ozone levels. The increase in mean ozone concentration with latitude and the existence

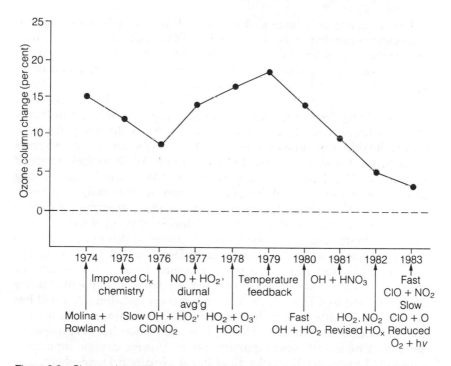

Figure 6.2   Changes in steady-state reductions in total column ozone for continuous releases of CFCs at 1975 rates as predicted from different models. Changes in the models and simulation techniques are indicated in chronological order
Source: US National Research Council, 1984

of seasonal cycles in ozone levels make it difficult to detect any trends in global ozone concentration. During the 1960s, ozone concentrations generally increased by up to 10 per cent at some monitoring stations (Komhyr et al., 1971), and this was believed to be associated with variation in solar radiation over the 11-year cycle (Keating, 1978). Analysis of ozone measurements for the 1970s, taken by 36 Dobson spectrophotometers throughout the world, reveal no evidence for either an upward or a downward trend, either on a global basis or a regional basis (Reinsel, 1981; Reinsel et al., 1981). Nevertheless, measurements taken by American satellites have pointed to a recent reduction of approximately 0.35 per cent per year which cannot be explained by variations in the solar cycle. In addition, the Antarctic lower stratosphere has shown a marked decrease in ozone concentration during the spring (September to November) since the late 1970s which may suggest that this region is particularly sensitive to the effects of the build-up of compounds depleting stratospheric ozone. Chlorine-containing compounds are generally held to be responsible for the creation of this 'ozone-hole' (Crutzen and Arnold, 1986; refer to Appendix 2).

That evidence of a large global ozone depletion is limited does not necessarily mean that ozone depletion by CFCs is not taking place: it may be that the current small decrease is being offset by natural effects or by the effects of other pollutants. Thus, for example, a global increase in atmospheric carbon dioxide concentration, through the 'greenhouse effect', may be warming the lower atmosphere but cooling the stratosphere, thereby shifting the equilibrium of stratospheric photochemistry in favour of ozone production (Groves and Tuck, 1979). A doubling of global carbon dioxide concentration is estimated to increase ozone concentration by between 3 and 6 per cent (Maugh, 1984; US National Research Council, 1984). However, the presence of CFCs, which have strong absorption bands in parts of the infra-red spectrum, may induce their own 'greenhouse effect' within the stratosphere, thereby offsetting the carbon dioxide effect (Ramanathan, 1975). Another way in which current reductions in stratospheric-ozone levels (maximum depletion expected near 40 km) may be offset has been suggested by Derwent (1982), who suggests that aircraft operations in the upper troposphere have caused an increase in ozone production which may have offset any depletion of the total ozone column caused by CFCs. He predicts that this cancellation effect will last into the 1990s, after which the effect of CFCs will dominate at least in mid-latitudes of the Northern Hemisphere. In the Southern Hemisphere, the impact of aircraft operations on the total ozone column amount is anticipated to be much smaller than in the Northern Hemisphere.

Since the early 1970s, CFCs have emerged as the compounds which are most likely to cause a significant depletion of stratospheric ozone in the near future – as may already be happening. Although early estimates of the depletion have been subsequently reduced, international concern for this pollution problem remains high. Whether this concern can be translated into effective international action remains to be seen and this is examined in chapter 11.

6.2   Carbon Dioxide

Carbon dioxide (CO_2) is a colourless trace gas which occurs naturally in the atmosphere, where it acts as an essential plant nutrient, and as an important determinant of the Earth-atmosphere thermal balance, so ultimately controlling our planet's weather and climate. Although virtually transparent to incoming short-wave solar radiation, carbon dioxide is a strong absorber of outgoing terrestrial radiation in the wavelengths of 7–14 μm. This in turn leads to energy which would otherwise escape to outer space being trapped within the atmosphere, warming the surface and lower atmosphere. Because the glass in a greenhouse traps the sun's heat (though it does this mainly by inhibiting convection, thereby stopping warm air rising and escaping), this process has come to be known as the 'greenhouse effect'. Several other trace gases such as methane, ammonia, chlorofluoro-carbons and nitrous oxide have similar thermal properties as carbon dioxide, and these gases are collectively referred to as 'greenhouse gases'.

Concern has been expressed that human activities, through the burning of carbon-based fossil fuels and changes in land-use practices, are increasing the global atmospheric carbon dioxide concentration such that this will result in a significant increase in surface air temperature and changes in other climate parameters. Many researchers predict that atmospheric carbon dioxide concentration will increase from the pre-industrial value of 260–300 ppm (the lower value gaining favour in recent research) to 600 ppm during the next century. Such a doubling of carbon dioxide concentration is estimated to cause an increase in the mean global surface air temperature of $3 \pm 1.5\,°C$ (US Environmental Protection Agency, 1983a; US National Research Council, 1979, 1982b, 1983b). The Earth would then be warmer than it has been for the past 125,000 years, the peak of the last interglacial, or possibly even warmer than it has been for the past two million years.

Accurate measurements of global atmospheric carbon dioxide concentration are available only since 1958 but they indicate a marked increase in carbon dioxide concentration since then. Between 1958 and 1980 carbon dioxide concentration at Mauna Loa Observatory, Hawaii increased by approximately 23 ppm to reach 338 ppm (figure 6.3). This corresponds to 48 Gt of carbon being added to the atmosphere during that period to produce a total weight of carbon in the atmosphere of 719 Gt (Clark et al., 1982). The pronounced seasonal variation in concentration

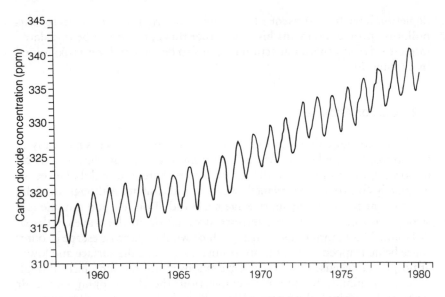

Figure 6.3 The atmospheric carbon dioxide concentration since 1958 as measured at Mauna Loa, Hawaii. By 1990 the concentration is expected to reach 352 ppm
Source: Holdgate et al., 1982a

of 5–6 ppm in figure 6.3 reflects an excess of photosynthesis over respiration during the spring and summer, and the reverse during the autumn and winter. Variations in sea-surface temperatures may be a minor contributor since the dissolved carbon dioxide content of the oceans is temperature-dependent (MacDonald, 1982). The seasonal cycle reaches 15 ppm at Point Barrow, Alaska but declines to 1.6 ppm at the South Pole. South Pole concentration of carbon dioxide lags behind that of Mauna Loa by a few ppm in any given year, consistent with the interpretation that most of the carbon dioxide is introduced in the Northern Hemisphere and that there is a time-lag in the mixing of the atmosphere between hemispheres.

To understand why carbon dioxide concentration has been increasing one must first examine the global carbon reservoirs and annual fluxes (figure 6.4). The fluxes between the reservoirs are relatively large, and compared with the other reservoirs or pools of carbon, the atmosphere is the smallest. Slight changes in the rates of exchange between reservoirs may bring about significant changes in atmospheric composition. The principal ways in which human activity directly affects the carbon content of the atmosphere is through fossil-fuel burning and changes in land-use practices. Human activities release carbon dioxide to the atmosphere through the burning of wood, coal, oil, natural gas, and other carbon-based materials. Annual global emission rates of carbon dioxide from

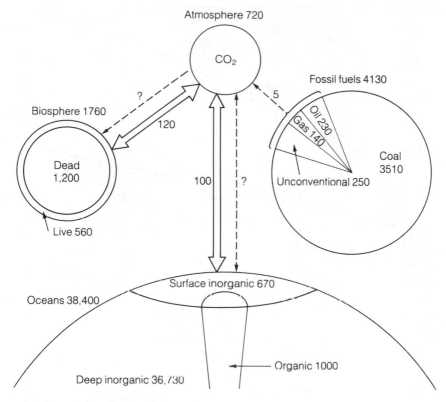

Figure 6.4 Global carbon pools and annual fluxes. Sizes of pools (circles) and annual fluxes (arrows) are shown in gigatons (10^{15}g) of carbon. Dashed arrows represent additional fluxes due to human activities
Source: Clark et al., 1982

fossil-fuel burning reached 5.3 Gt of carbon per year in 1980 (Clark et al., 1982). The cumulative total release since 1860 is approximately 160 Gt of carbon. Annual carbon dioxide emissions have varied considerably since 1860, with the post-war average rate of approximately 4.5 per cent per year having been reduced today to less than 2.5 per cent per year because of the emergence of the energy crisis in oil and gas in 1973 and the associated emphasis on energy conservation (figure 6.5). Estimates of future rates of growth are of critical importance for predicting future concentration of atmospheric carbon dioxide. However, future emission rates are difficult to predict as they will be influenced by rates of population and economic growth, by improvements in energy efficiency, and by the changing mix of energy supply (figure 6.6).

A net release of carbon dioxide to the atmosphere occurs whenever changes in land-use result in ecosystems of high carbon density (e.g. forests)

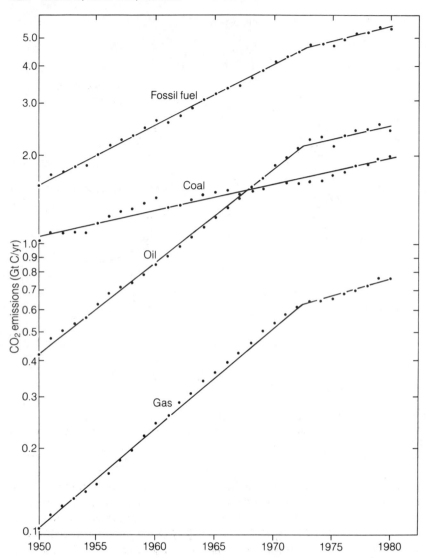

Figure 6.5   Annual global carbon dioxide emissions from various fossil fuels for the period 1959-80 highlighting the effects of the energy crisis in 1973
Source: Clark et al., 1982

being replaced by those of a lower carbon density (e.g. agricultural or grazing land). Forest harvesting can also be a contributory factor if its rate exceeds that of forest re-growth (e.g. deforestation in the Amazonian basin). Disturbed soils are an additional major source of atmospheric carbon. Some changes in land-use can provide a sink for carbon dioxide – for

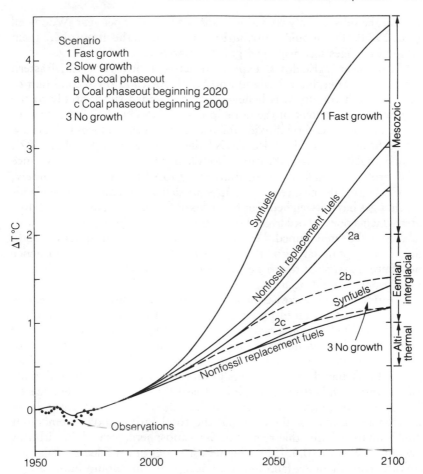

Figure 6.6 Projections of global temperature change caused by various economic-growth and fuel-mix scenarios
Source: Hansen et al., 1981

example, when farmland is allowed to reforest or afforestation programmes are pursued. In addition, those plants for which carbon is a limiting nutrient should experience enhanced growth with increasing carbon dioxide in the atmosphere (what is termed the carbon fertilization or β-effect) which in turn will increase the biosphere carbon sink. Estimates of carbon releases from the bisophere to the atmosphere in response to changes in land-use are disputed and range from a net source of 8 Gt of carbon per year to a net sink of 2 Gt per year (World Meteorological Organization, 1981). Revisions of earlier estimates have frequently been made: for example, whereas in 1978 Woodwell estimated 8 Gt of carbon per year, he subsequently revised his estimate to 2.6 Gt per year (Woodwell et al., 1978). The annual contribution of carbon to the atmosphere from land-use changes has displayed great variability.

The 'pioneer' agricultural explosion across North America, Eastern Europe, Australia, New Zealand and South Africa in the second half of the nineteenth century was believed to have added substantially to the carbon dioxide content of the atmosphere (Adams et al., 1977; Wilson, 1978). However, carbon dioxide release arising from changes in land-use during the past two or three decades is believed to be small (Revelle, 1982). Total cumulative release of carbon dioxide from changes in land-use since 1880 probably equals the contribution by fossil-fuel burning – namely, 160 Gt. During the past century, it appears that carbon dioxide from land-use changes has outweighed that from fossil-fuel burning until the 1940s, after which a reversal occurred. Future predictions suggest that this relative balance will be maintained. Furthermore, increasing atmospheric carbon dioxide may stimulate growth in photosynthesis production and hence biosphere mass, such that a net flux of carbon from the atmosphere to the biosphere could occur.

To estimate future increases of atmospheric carbon dioxide it is necessary to know what fraction of future human net carbon dioxide emissions – from fossil fuel burning and land-use changes – will remain in the atmosphere. The oceans have long been assumed to absorb one-half of the carbon dioxide added to the atmosphere by human activities (calculations based on measured global atmospheric carbon dioxide concentration and estimates of global fossil-fuel carbon dioxide release). However, taking into account carbon dioxide release as a result of land-use changes, it may be the case that the total carbon dioxide emissions from human activities that remain in the atmosphere, termed the airborne fraction, is closer to 0.4. If even higher estimates of net releases of carbon dioxide from changing land-use were accepted, this would imply a very low airborne fraction, of perhaps 0.2. In other words, the concern that in the past land-use changes have released large quantities of carbon dioxide is largely offset by the lower airborne fraction that is implied by such higher estimates. Using an estimate of 0.4 for the airborne fraction and the current energy growth rate of 2.5 per cent per year, Clark et al. (1982) estimate that a carbon dioxide concentration of 600 ppm would be reached by the year 2080. This is similar to the US Environmental Protection Agency (1983a) and US National Research Council (1983b) who suggest that 600 ppm of carbon dioxide will be reached between the years 2060 and 2080.

The oceans play a crucial part in the carbon dioxide issue yet there exists only limited understanding of oceanic processes. The upper 70–100 m of the ocean is heated by the sun and agitated by the wind such that it is relatively well mixed and is an effective absorber of carbon dioxide from the atmosphere. Beneath this is a stagnant region, the thermocline, stabilized by decreasing temperature and increasing density to a depth of about 1000 m. Below this is the much larger region of cold ($< 5°C$) deep ocean, isolated from the surface waters by the thermocline (Baes et al., 1977). The downward diffusion, advection or convection of carbon dioxide-containing waters of the mixed layer into the deeper water is a relatively slow process taking 500 or more years (Revelle, 1982). Although the oceans have the potential for absorbing all excess atmospheric carbon dioxide, in practice the slow transfer between the surface and deep ocean prevents realization of its potential as a carbon dioxide sink – at least in the time-scale with which we are concerned. Furthermore, as the upper oceanic layers become saturated with carbon dioxide, the fraction of the atmospheric carbon dioxide which the oceans can absorb may be reduced, thereby exacerbating the carbon dioxide problem.

6.2.1   Effects on Climate

There is a large number of simple and complex models which attempt to quantify the effects on climate of an atmosphere with a considerably higher carbon dioxide concentration. The predictions usually refer to the globally and annually averaged surface air temperature difference between a pre-industrial world with 300 ppm of carbon dioxide and a future world with a carbon dioxide concentration of 600 ppm. The majority of the models predict a global warming of $3 \pm 1.5°C$. A few researchers such as Idso (1980, 1984) predict negligible temperature changes but their models have been criticized for omitting key positive feedbacks (US National Research Council, 1982b).

Because of positive feedbacks such as the melting of snow and ice which decrease the surface albedo, and the increased evaporation leading to an enhanced water vapour greenhouse effect, the temperature increase at high latitudes for a doubling of the carbon dioxide concentration should be two or three times greater than the warming experienced in the tropics where enhanced convection, carrying sensible heat away from the ground, acts as a negative feedback. The US National Research Council (1979) suggests that lower latitudes will experience an annual mean temperature increase between 1.5 and 3.0°C while the polar and subpolar latitudes will experience a warming of 4.0–8.0°C. Parkinson and Kellogg (1979) cite temperature increases as high as 10°C for the polar latitudes and less than 2°C for the low latitudes due to the effect of feedbacks.

The greater response of high latitudes to the carbon dioxide greenhouse effect suggests that a carbon dioxide-induced warming may be first detected in those regions. Even so, only when a sustained global warming trend is evident will the carbon dioxide greenhouse effect be generally accepted. Given that the Earth's climate is controlled by many factors, it is necessary to separate out their relative importance so as to reveal the change in climate resulting from increasing carbon dioxide. Such an approach is necessary because a warming trend due to carbon dioxide changes may initially be counteracted by a cooling trend caused by other factors (Broecker,1975). Major advances have been made in this direction of research in recent years: for example, Gilliland (1982) explained 87 per cent of the Northern Hemisphere 1881–1975 temperature variance (for time-scales greater than three years) by forcing due to carbon dioxide, volcanoes and the 76-year solar cycle. This model confirms the importance of the carbon dioxide forcing of the climate, but it suggests that a doubling of carbon dioxide concentration will produce an increase in surface air temperature of $1.6 \pm 0.3^{\circ}C$, which places it in the lower range of the generally accepted predictions of temperature increase (Gilliland and Schneider, 1984). Using an alternative model, Hansen et al. (1981) concluded that radiative forcing by carbon dioxide together with volcanoes accounted for 75 per cent of the variance in the 5-year smoothed global temperature for the past century, while the addition of a hypothesized variation in solar luminosity increased the figure to 90 per cent. Hansen et al. (1981, 1983) believe that the carbon dioxide greenhouse warming should emerge from the noise level of natural climate variability by the end of the century, and there is a high probability of warming in the 1980s. Indeed, the Northern Hemisphere mean temperature decrease which began in the 1940s has been arrested and even reversed during the mid-1970s (figure 6.7). However, other climatic influences, such as volcanic eruptions, may yet counteract and obscure a carbon dioxide-induced warming

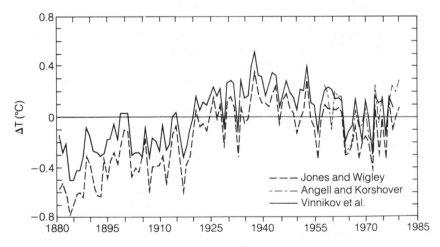

Figure 6.7   Annual surface temperature anomalies for the Northern Hemisphere, 1881–1980, according to various researchers
Source: Clark et al., 1982

trend such that undisputed confirmation of its detection may not be available for some time.

Of more practical importance than estimates of global annual average temperature increase is that of resulting regional and seasonal changes. Similarly, although a warmer Earth implies increased evaporation and precipitation, it is important to know how these two components of the hydrological cycle will balance in particular regions and seasons. Model predictions of regional changes in temperature and moisture in a future world with 600 ppm of atmospheric carbon dioxide have to be treated with much caution. This explains why attention has been given to producing future warm-Earth scenarios based on past climatic data. Wigley et al. (1980) provide such a Northern Hemisphere scenario by comparing the five warmest years with the five coldest over the period 1925–74. The average temperature difference between the warm and cold year groups was 1.6°C for high latitudes and 0.6°C for the Northern Hemisphere as a whole – that is, considerably less than the temperature difference predicted for a doubling of atmospheric carbon dioxide content. In their analysis, maximum warming occurred in winter in high latitudes and in continental interiors. In contrast, some regions such as Japan and much of India showed negative temperature differences (figure 6.8a). An examination of the pressure pattern differences between cold and warm years showed that the warm years experienced intensified high latitude (50–70°N) westerlies, greater cyclonic activity in the Arctic and subarctic regions of the Eastern Hemisphere, and a westward displacement of the Siberian high during winter. Precipitation changes showed only a slight (1–2 per cent) overall increase in warm years compared with cool ones. Precipitation decreases occurred over much of the United States, most of Europe and Russia and over Japan, while increases in precipitation over India and the Middle East were indicative of a more intense monsoon circulation in warm years (figure 6.8b).

Another approach to presenting a warmer Earth scenario, based on past climatic data, has been to reconstruct the climate of the mid-Holocene Altithermal or Hypsithermal period of 8000–4500 years ago when temperatures were 1.5–2.5°C higher than today in middle latitudes. Butzer (1980) highlights the fact that this was a period when drier conditions prevailed in the wheat- and corn-producing belts of North America and the Soviet Union, while moderately increased precipitation occurred in the tropical and subtropical arid zones. Kellogg and Schware (1981) derived possible soil moisture patterns for a warmer Earth based not only on reconstruction of the Altithermal period, but also on a comparison of recent warm and cold years in the Northern Hemisphere, and on a climate model experiment. Their analysis pointed to wetter conditions in subtropical continents but drier conditions in the central United States (figure 6.9). In contrast to Wigley et al. (1980) and Butzer (1980), Kellogg

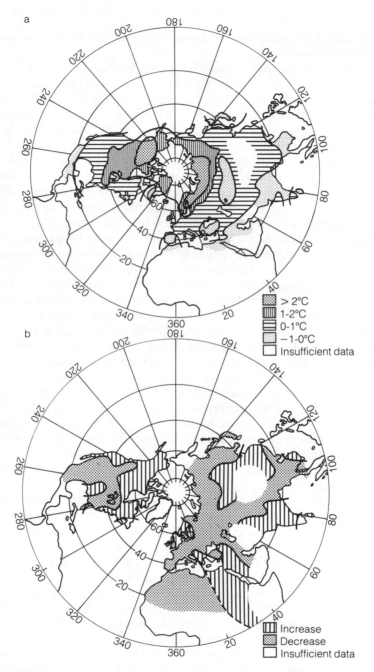

Figure 6.8a Mean annual surface temperature changes from cold to warm years
Figure 6.8b Mean annual precipitation changes from cold to warm years
Source: Wigley et al., 1980

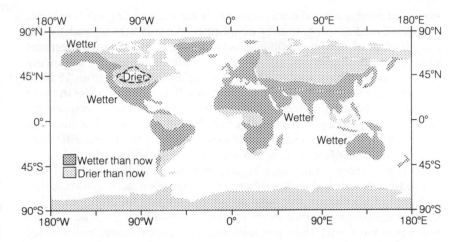

Figure 6.9   A climate scenario in which deviations from present growing-season soil-moisture patterns are plotted for a warmer Earth
Source: Kellogg and Schware, 1981

and Schware (1981) as well as Kellogg (1978) point to wetter conditions in Europe and in the southern Soviet Union. Such conflicting predictions highlight the limitations of the analogies and models currently employed.

One of the principal concerns emerging from scenarios of regional climates which may be experienced due to a carbon dioxide-induced warmer Earth is that of warmer and drier conditions in the Corn Belt of the United States, especially since much of the world food market depends upon the availability of grain from that country. Bach (1978, 1979) studied data on weather and corn yield in the United States Corn Belt states from 1901–72 and estimated that corn production in those states would change by about 11 per cent for every 1°C change in average maximum temperature over the growing season, and by about 15 per cent for each 10 per cent change in precipitation. Bach concluded that warmer and drier weather would decrease corn production in the Corn Belt, and cooler and wetter weather would increase it. In terms of optimum climatic conditions, the Corn Belt would be forced towards the north onto acid podzol soils which are badly leached and which would require extensive and expensive soil amendments. Any reduction in North American grain yields has potentially severe consequences for the world grain market in terms of food shortages, especially for Third World nations with poor purchasing-power.

However, Rosenberg (1982) has questioned the validity of estimates of the impact of climatic change on crop yields which assume that crop varieties and production technologies do not change over time. He quotes the example of hard red winter wheat which has already considerably expanded northward and northwestward since 1920 – being made possible by plant-breeding programmes of more cold-hardy strains and by the introduction of new technological advances and dry land-management practices such as summer fallow and stubble mulching. In advancing northward from Nebraska to Manitoba during the period from 1920 to 1980, the hard red winter wheat adapted to a 20 per cent decrease in mean annual precipitation, a 4.2°C decrease in the mean annual temperature and a 10-day decrease in the length of the frost-free season. The southern movement of wheat in Texas also involved a significant climatic difference of a 12 per cent decrease in mean annual precipitation, a 1.6°C lower mean annual temperature and a 10-day increase in the length of the frost-free season. Not only does Rosenberg (1982) question the implication that a change in US climate will have an adverse impact on crop yields, but he also questions whether evapotranspiration, which may be a more important parameter than precipitation and temperature, will change adversely. Rosenberg suggests that evapotranspiration in much of the grain belt may be unchanged or even beneficially reduced, despite changes in other climatic parameters.

6.2.2   Direct Effects on Crop Yields

Whether the changes in climate will adversely affect crop yields in major grain-producing areas of the world may be debated, but there is an increasing consensus that increased atmospheric carbon dioxide will directly enhance crop yields. Controlled growth chamber experiments (as well as greenhouse practices) have established beyond doubt that many plants for which carbon is a limiting nutrient respond to short-term carbon dioxide enrichment with faster growth and greater yields (Rosenberg, 1981). To what extent this applies to field conditions and to the unmanaged biosphere is as yet unclear.

At full outdoor light intensity, net photosynthesis in many plants increases with carbon dioxide concentration up to at least 900 ppm (Cooper, 1982). The extent and nature of the effect depends upon plant biochemistry, growth form, age of plant, and status of water and nutrients (phosphorous and nitrogen) in the soil, among other factors. The two broad categories of plants designated C_3 and C_4 differ in the biochemical pathways through which they fix carbon dioxide in photosynthesis and so differ in the degree to which they benefit from increased carbon dioxide. At present atmospheric levels of carbon dioxide, C_4 plants have substantially higher rates of photosynthesis than C_3 plants at all temperatures above approximately 15°C. As carbon dioxide concentration increases, C_3 plants benefit more until at 1000 ppm, some C_3 plants perform almost identically to comparable C_4 species. C_3 species include temperate species such as small grains, legumes, most grasses, and most forest species, while C_4 species are generally those adapted to warm climates and include tropical grasses, corn, sorghum, maize and sugarcane.

Carbon dioxide fertilization may also benefit plants by increasing their efficiency in water use. In some plants, carbon dioxide induces partial closure of stomates through which carbon dioxide enters the leaf for photosynthesis and through which water vapour simultaneously escapes in transpiration. This partial closure should still allow the same amount of carbon dioxide to enter, as the carbon dioxide gradient would be greater in a higher carbon dioxide environment, but should reduce the loss of water vapour as the humidity gradient will be little altered. The net effect is that some plants should be more resistant to water stress (and more tolerant of atmospheric pollution as partially closed stomates impede entry of potentially harmful air pollutants into leaves). Sionit et al. (1981) found that carbon dioxide fertilization completely compensated for lack of water when wheat was grown in the laboratory without enough water for maximum development. Whilst water stress periods depressed production of wheat grown under normal carbon dioxide levels, wheat grown under 1000 ppm of carbon dioxide produced as much as unstressed plants grown at the current atmospheric levels of carbon dioxide, and remained turgid at moisture levels that wilted their unexposed counterparts.

Kimball (1982) reviewed 437 experiments published in this century which dealt with yields of agricultural crops exposed to varying carbon dioxide concentrations, and concluded that a doubling of atmospheric carbon dioxide could increase global productivity by 33 per cent without additional inputs of water or fertilizers. Other researchers express more cautious optimism as they point out that the carbon dioxide fertilization effect may be offset in some regions by changes in climate. Revelle and Shapero (1978), for example, point out that if average cloudiness increases, the quantity of incoming solar radiation will be lowered, and the energy available to crop plants for photosynthesis will diminish. However, whether changes in climate will offset the crop-yield gains attributable to direct enhancement of photosynthesis by higher levels of carbon dioxide will continue to be debated (Waggoner, 1984).

Although undisputed confirmation of a global warming trend due to the carbon dioxide greenhouse effect is not yet available, and the benefits or disadvantages of higher atmospheric carbon dioxide concentration and the predicted climate changes are still being debated, many authorities are urging that international action be taken on the carbon dioxide issue. Clearly, increasing atmospheric carbon dioxide and its potential effects on climate and crops is a global issue, and if effective action is to be taken to counter any unwelcome effects, it requires all nations to co-operate in a pollution-control strategy (Hare, 1982). The issue is likely to be a stern test of whether pollution control can be tackled rationally and successfully at an international level (refer to chapter 11).

Having reviewed the nature, sources and effects of atmospheric pollution in the foregoing chapters, we now move on in Part II to examine in detail the various national and international approaches to atmospheric pollution control.

Part II

National and International Approaches to Atmospheric Pollution Control

7 Air-pollution Control Strategies

7.1 Introduction

An air-pollution control strategy refers to the master plan adopted by a country, or sometimes a group of countries (e.g. the European Community) to tackle air pollution problems and to ensure that air-pollution concentrations are reduced or are maintained below a specific or general level that is deemed acceptable. Strategies may be short-term or long-term. They may differ from country to country, as may the level of pollution deemed acceptable, but there are common elements which make it possible to distinguish four types of strategies: (1) air-quality management, (2) emission standards, (3) economic, and (4) cost-benefit. Although all four strategies may be separated so as to discuss their major characteristics, in practice countries commonly adopt a combination of two or more types. This is illustrated in later chapters which examine selected national approaches to air-pollution control in detail.

7.2 Air-quality Management Strategy

The Air-quality Management strategy, sometimes referred to as the air resource management strategy, involves designating the level of pollution deemed acceptable in terms of a set of ambient (outdoor) air-quality standards and then controlling pollutant emissions to ensure that these legal limits are not exceeded (Weber, 1982). de Nevers et al. (1977) define air-quality management as the 'regulation of the amount, location, and time of pollutant emissions to achieve some clearly defined set of ambient air-quality standards or goals' (figure 7.1). Obviously an approach which involves regulating emissions from millions of stationary and mobile pollution sources which have varying characteristics, locations and use patterns is not simple and the strategy produces a complex set of control regulations. Nevertheless, it is an approach employed by many countries: it was first used by the Soviet Union in 1951 but it is the United States to

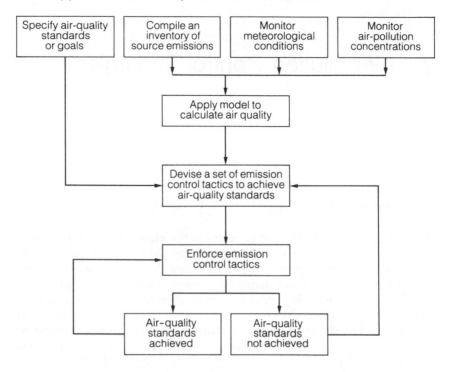

Figure 7.1   Stages involved in the Air-quality Management strategy

which this strategy is most associated and which has made known its advantages and disadvantages. The strategy involves several steps which can usefully be discussed within a five-fold typology.

Firstly, ambient air-quality standards or goals need to be specified. These standards may vary from country to country and they may be changed from time to time within a given country. Standards may be applied nationally, regionally or according to local conditions of land use, topography, meteorology, and so on. Standards are usually based on detailed scientific investigations of the effects of various levels of individual pollutants (or combinations of pollutants) on public health, animals, vegetation, materials, etc. The standards also involve economic, social, technical and political considerations. Given the inclusion of so many considerations, it is not surprising that standards may be amended from time to time: for example, standards may be made more stringent if new scientific research reveals a lower threshold of effect than was previously believed to exist. Standards may be relaxed if, say, the cost of controlling pollutant emissions in order to achieve the standards is deemed to be economically unacceptable to a nation.

Secondly, to know if the actual pollution concentrations are in compliance with the standards, an air-pollution monitoring system is required. This requires adequate numbers and distribution of accurate monitoring equipment and specialist staff to maintain and operate the equipment. It is vital to have accurate knowledge of existing air quality, yet in practice, problems arise. Thus, for example, although the United States has an impressive air-quality monitoring network involving over 9400 stations, some doubts have been raised concerning the effectiveness of this system. The US General Accounting Office (1979) sampled 243 monitoring stations and revealed that 81 per cent had one or more problems that could adversely affect the accuracy and reliability of air-quality data.

Thirdly, in order eventually to control the pollutant emissions in an area so as to achieve the air-quality standards, an inventory of current emissions from the various sources is needed. Ideally, the emission rates of pollutants and plume-rise parameters should be known for each hour of the day and day of the year for all sources of pollution. In practice, only approximate figures can be obtained, except for the sources which are major individual contributors to the overall emissions in an area and therefore subject to more precise measurement. In the case of the large number of small emission sources, emission inventories frequently have to resort to treating them collectively as area sources.

Fourthly, a mathematical model is needed to be able to calculate air pollution levels from an emission inventory of an area under varying meteorological conditions and to predict air-pollution levels arising from proposed changes to emissions in an area. Such a model is central to the management of air quality in order that alternative regulations of emissions can be examined for compliance with air-quality standards. The accuracy of models varies considerably because of the ability with which various models can handle the emission information and the complexities of physical and chemical relationships and atmospheric dispersion. The US National Commission on Air Quality (1981) typically found that models may overpredict or underpredict from actual concentrations by a factor of two.

The simple 'proportional' or 'rollback' model assumes linearity in the relationship between emission rates and resulting pollution concentrations such that, for example, a region with sulphur dioxide levels at twice the air-quality standard will attain the standard if total sulphur dioxide emissions in the region are halved. Given that this model has been so widely adopted, de Nevers et al. (1977) outline in detail the many limitations of this model. Of greater sophistication are the diffusion or dispersion models which can predict pollution concentrations from emission invent-ories under varying meteorological conditions and take the topography of the area into account. Such models not only allow various emission

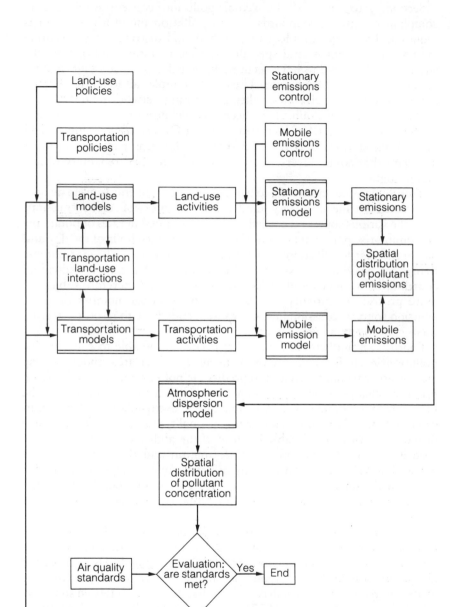

Figure 7.2   Air-quality simulation system
Source: Gross, 1982

control programmes to be examined for existing pollution sources, but they can also assess the likely impact of new pollution sources. More elaborate models as outlined by Gross (1982) are able to assess the consequences of proposed land-use and transportation policies (figure 7.2). In this model, modelling land-use and transportation policies represents the first stage of this Air Quality Simulation System (Gross, 1982). The output from this stage is passed to the emission stage of the model where, based on emission characteristics, the transport and land-use activities are converted to pollution emission rates. The output from this stage represents the geographical distribution and intensity of emissions over the region. Based on this information, meteorological and atmospheric dispersion models are employed to predict the spatial distribution of pollution concentrations over the area. Such a map of predicted pollution concentrations reveals whether the proposed emission-control programme will result in compliance with the air-quality standard for a particular pollutant or group of pollutants. Should it reveal areas which would fail to meet the standard, then alternative sets of options for emission control can be considered.

Fifth and finally, given the need to reduce emissions in an area in order to achieve compliance with air-quality standards, a set of emission-control options or tactics is needed. The more emission-control options available to a manager, the more flexibile the air-quality management strategy can be made to suit a given situation. At one extreme, emission-control options may involve comprehensive integral planning of energy supply, transportation, land-use, and industrial development for a region. At a more specific level, emission-control options may include the setting of emission standards, restrictions on the use of certain fuels, the requirement that emission-control equipment be fitted, the closing down of factories, and the adoption of tall stacks. All such options need to be examined for their effectiveness in reducing emissions so as to achieve air-quality standards, as well as for their technical and economic feasibility, the speed with which they can be implemented, and their enforceability.

One major criticism of the air-quality management strategy relates to its use of air-quality standards. Although such legal standards result in pollution concentrations being reduced to an acceptable level in a 'polluted' area, pollution in a 'clean' area is allowed to increase up to the standard. This entails degradation of air quality where it is now high. To overcome this limitation, many countries adopt a special 'nondegradation' or 'no significant deterioration' policy to prevent increased pollutant emissions in clean areas such as wilderness areas, national parks, scenic locations, and so on. Another criticism of this strategy is that it may result in widely varying control requirements and pollution-control expenditures for competing companies in regions of differing air quality. Stringent control may be placed on an industrial plant in an area of non-compliance while an identical plant in an area of compliance may be subject to little or no control. However, in order to reduce such anomalies, which could result in unfair economic competition, emission design or performance standards for particular industrial plants or processes are usually applied nationally. Such action represents the adoption of aspects of a second air-pollution control strategy – namely, the emission-standards strategy. It is not unusual to find that countries adopt the best parts of both strategies in order to tackle air-pollution problems. The United States, for example, uses the air-quality management strategy for the control of pollutants called 'criteria pollutants' (for which air-quality standards have been set) and employs the emission-standards strategy for another list of pollutants termed 'hazardous pollutants' (e.g. beryllium, mercury, asbestos).

7.3   Emission-standards Strategy

The Emission-standards strategy specifies the maximum amount or concentration of a pollutant which is allowed to be emitted from a given source. Emission standards are designated for a large number of pollutants or combination of pollutants and may be applied to individual or specific groups of emitters. If emission standards are derived from consideration of air-quality standards this makes the emission-standards strategy really part of an air-quality management strategy. If emission standards are derived from consideration of the best available control technology and from economic considerations, this strategy is entirely independent of the air-quality management strategy. Frequently, this strategy is referred to as the 'best practicable means' or 'good practice' strategy and is associated with the traditional British approach to air-pollution control. The term 'practicable' implies 'best available and economically feasible control technology'. Such an approach attempts to reduce air pollution to the greatest extent possible with the practical methods available, but that the cost of doing so should not be unreasonable (World Health Organization, 1972).

   Emission standards come in a variety of forms, such as the maximum concentration of a substance in a given volume of gaseous effluent or the maximum opacity of a smoke plume. The former type of emission standard may be qualified further by specifying the type of control devices that sources should install in order to fulfil the emission standards (i.e. equipment or design standards). In other cases an emission standard may control the nature of the fuel burned: for example, in the case of sulphur dioxide emissions, an emission standard to limit the sulphur dioxide emitted per unit of fuel burned specifies the use of coal with a low sulphur content – say, 1 per cent by weight. Similar standards may be applied to lead in petrol. Some flexibility is possible with some emission standards such that, for example, emission standards may be adjusted in relation to stack height, in recognition of the better dispersion achieved by higher stacks.

Since it is more expensive to install pollution-control devices in existing installations than in new ones, this strategy often leads in practice to the specification of different requirements for new and existing units. As technology improves, thereby making it possible to set a lower emission standard, new sources face increasingly more strict controls.

The emission-standards strategy is a relatively simple strategy to apply and as such it is a strategy that many countries initially adopt in order to control pollution from large single-emission sources which are deemed to be major polluters. Some argue that the emission-standards strategy is better than the air-quality management strategy, since once air-quality standards are set, one has to calculate the emission reduction required of sources. The best that can be done in trying to meet the air-quality standards is to persuade each emitter to reduce emissions to meet some emission standard, which will ultimately be based on someone's judgement of what is 'good practice'. Thus, they argue, why not short-circuit the development of ambient air-quality standards, air monitoring, diffusion modelling, etc. and simply require all emitters to go directly to a 'good practice' emission standard? However, this view highlights the major criticism of the emission-standards strategy in that it does not make publicly evident any explicit judgement as to what level of air quality is to be regarded as acceptable, whether for the protection of public health, vegetation, animals, materials, or amenity. It does not guarantee, however comprehensive it may be, the achievement or maintenance of acceptable air quality (Wilde, 1978). In practice, however, the decision as to what emission standard is to be adopted inevitably considers not only economic and technological aspects but the effects of air pollution on human health and the environment, as well as the political climate and public opinion (World Health Organization, 1972). Indirectly, therefore, air quality is frequently considered to some extent in this strategy and it is seldom found in its pure form.

7.4 Economic Strategies

Ideally, strategies using an economics-based approach to air-pollution control would provide financial incentives for emission sources to pursue the most cost-effective means for reducing pollution (Anderson et al., 1977; Magat, 1982; Rosencranz, 1981). A variety of economic strategies may be used to control air pollution, although the use of emission charges and pollution permits are the most frequently cited.

The emission-charges strategy is based on requiring polluters to pay charges related to the amount of pollution they emit, the purpose being to leave polluters the choice to pollute or pay. This strategy recognizes that different pollution sources have different marginal costs of control, that some can control more cheaply than others, and that there would be net savings to society if some plants reduced emissions to a greater and others to a lesser extent. A properly set uniform emission fee that charged a given sum for each unit (say, 1 tonne) of pollutant emitted, for example, could achieve a collectively established level of air quality at the least total control cost to society (Krier and Ursin, 1977). Assuming that each source wishes to minimize its total costs under the emission-charges system, each would blend control expenditure (to reduce its emissions and thus its emission-charge liability) and emission-charge payments (on those units of pollutants emitted after control) in the way it finds cheapest. Such an economic approach provides incentives to firms to find new techniques and new products which lead to lower pollutant emissions.

The air quality which results from this strategy depends upon ensuring that the charges are set at precisely the right level for all pollutants. The emission charge may have to be continually adjusted until the air quality which society deems acceptable has been achieved. In this sense, this strategy may be viewed as a tactic employed in the air-quality management strategy to meet air-quality standards. However, it could be employed in its pure form and, perhaps the fees used to compensate victims of pollution and offset damage caused by the pollution. Thus, for example, several international airports have set up a system of charges related to the level of aircraft noise, with the funds raised being used to finance the insulation of nearby homes. The emission charges for each pollutant would have to be adjusted to take into account increases in the number of emission sources (if air quality is not to deteriorate) and it would have to be adjusted to reflect increases in general price-levels. Setting the fees for each pollutant would be highly contentious and beset by political wrangling.

This strategy has its limitations in that polluters with strong market power could simply pay the charges and pass them on to consumers. Also, the equipment to monitor continuously and precisely pollutants at their source is in many instances not available. Since this strategy requires monitoring emissions from individual sources, it is a strategy which tends to be limited to the larger industrial emission sources. This is the form in which it has been adopted in Czechoslovakia and East Germany. To apply this strategy to numerous small-scale emission sources such as heating in homes and motor vehicles would be more complex. Even so, Mills and White (1978) have proposed an 'effluent fee' for motor-vehicle emissions calculated by measuring emission levels at the beginning and end of the year multiplied by the distance driven. They argue that this effluent fee would give motorists the proper incentive to ensure that exhaust emission-control devices were efficiently maintained.

There are many forms of emission charges which may be used to provide economic incentives to polluters either to reduce or cease their pollution, including sales taxes on fuels or fuel ingredients: for example, Norway introduced fuel taxes in 1971 whereby a basic tax is levied on each unit of oil, together with an additional charge based on its sulphur content. To provide a further economic incentive for emission sources to reduce the pollutant emissions, Norway later introduced rebates to polluters in proportion to the amount of pollution reduction achieved by them (OECD, 1980c). Penalty fees or fines constitute another form of economic incentive for not violating emission standards and these are widely adopted. Suzhou, for example, a city in the Jiangsu Province of China, assesses the fine to be paid by a violator as the volume of wastes emitted (in tonnes or cubic metres) multiplied by the amount of pollutants exceeding the national standards (in percentage terms), multiplied by the standard fee for each unit of waste which exceeds the national standard (Chang, 1985). In Japan, a fee is charged based on emission levels for large sulphur dioxide sources in polluted areas. The proceeds are used in part for the medical care of patients affected by air pollution (Wetstone and Rosencranz, 1983). Other economic incentives adopted by some countries include subsidies or grants for the installation of pollution-control equipment, tax remissions for investment in or operation of pollution-control equipment, tax remissions for industries which use wastes as their main raw material for production, and import duties and quotas on pollution-producing equipment or materials (de Nevers et al., 1977). The choice of economic incentive ultimately depends upon the purpose of this strategy, whether it is to reduce specific types of pollution generally or in given geographical areas, to reallocate funds for investment in pollution control or to compensate victims of air pollution (Rosencranz, 1981).

An alternative economics-based or market-oriented strategy which avoids some of the problems of the emission-charges strategy is a system of 'pollution permits' or 'pollution rights'. The regulating authority determines the total amount of emissions to be allowed in an area and then issues the equivalent number of permits or rights to pollute, which can then be marketed, auctioned or allocated in some way. Although such a strategy is possible as a distinct strategy, it is more often employed as a tactic for reducing emissions as part of the air-quality management strategy.

In the United States, an 'emissions trading policy' has been adopted which represents a move in the direction of a marketable pollution permit system (Brady, 1983). The emissions trading policy established in 1982 incorporates previous policies known as 'bubble', 'offsets' and 'banking' into a single transactional instrument, the 'emissions-reduction credit'. The bubble policy was the first implicit market-based approach to be used in reducing emissions in the United States. It allows for firms to place an imaginary bubble over the multiple point sources of its plant and to be

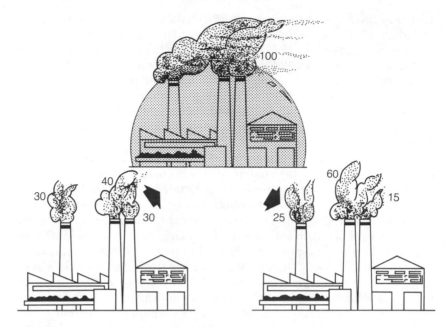

Figure 7.3 The 'bubble' concept: pollutants of a like kind from each source within a plant can be controlled to minimize costs, providing the total emission from the plant stays within a set limit
Source: supplied by US Environmental Protection Agency, Research Triangle Park, North Carolina

given an overall maximum emission limitation for the bubble (figure 7.3). Within the bubble, firms are allowed to increase emissions at one point-source as long as greater or compensatory reductions in emissions are accomplished at another point source within the bubble. Such a policy allows for intrafirm trades in emission reductions. The 'offset' policy extends the bubble concept so as to allow interfirm trading of emission permits among activities not located in the same plant or not owned by the same firm. It requires a greater than one-for-one reduction in emissions in order to achieve a net improvement in ambient air quality. Firms which reduce emissions by amounts exceeding the reductions required by the authorities under the bubble or offset policies can 'bank' emission-reduction credits and either use them to expand their own plant later or openly trade them with other companies. Obviously, constraints have to be imposed on transactions in emission-reduction credits in order to ensure that trades do not contribute to a violation of the air-quality standards (Brady, 1983).

Given the success of this system to date, it is likely that firms will press for much wider adoption of this strategy for pollution control in the United States (refer to section 8.11). However, it requires accurate monitoring of emissions and air quality, as well as an effective supervision and enforcement system for this strategy to be successful. Of course, with this strategy being market-oriented, it is possible that one firm could purchase more pollution permits than it needed, with the intention of keeping out competitors, and the market would cease to function. Interestingly, Rosencranz (1981) points out that conservation groups could conceivably buy up all available permits to ensure improved air quality. However, this is unlikely to occur given that the whole concept of the buying and selling of the right to pollute is unacceptable to most environmentalists.

7.5   Cost-benefit Strategy

A cost-benefit strategy first attempts to quantify the costs of all of the damage resulting from air pollutants and the costs of all known ways of controlling those pollutants, and then adopts the pollution-control option(s) which minimizes the sum of pollution damage and pollution-control costs. Figure 7.4 illustrates this approach in its conceptual form. If no pollution-control expenditures are made, ambient air-pollution concentration will be high and pollution-damage costs high. As control expenditure increases, the pollution concentration and associated damage costs will fall. Expenditure costs rise very steeply as the ambient concentration approaches zero. The damage cost curve is shown starting at some low value – at a small or zero concentration – and increasing rapidly at high concentrations. The sum of the two is shown to have a minimum value at some intermediate concentration. This minimum is the optimum; expenditure above or below it is deemed wasteful for society as a whole (de Nevers, 1981).

Such an approach is exceedingly complex, given that one must assign values for all kinds of pollution damage including premature death, varying degrees of illness, property deterioration, reductions in yield and quality of crops, damage to irreplaceable historic monuments, and visibility degradation. All the effects of pollutants need to be considered for both short- and long-term exposure. Similarly, all the costs of pollution-control options must be considered, including equipment, administration, effects on patterns of development, and unemployment. At present our knowledge is too inadequate to allow us to quantify all of the many variables involved in cost-benefit analyses of pollution effects and pollution control. Nevertheless, some progress is being made in assessing the costs and benefits associated with even the most complex pollution problems, such as visibility degradation. Kneese (1984) describes a methodology being developed to determine how much people are willing to spend to preserve good visibility conditions in National Park areas. In the future, the cost-benefit strategy will gain increased favour with nations attempting to integrate pollution control with other policies such as agriculture, industry, energy, transportation, and land-use management.

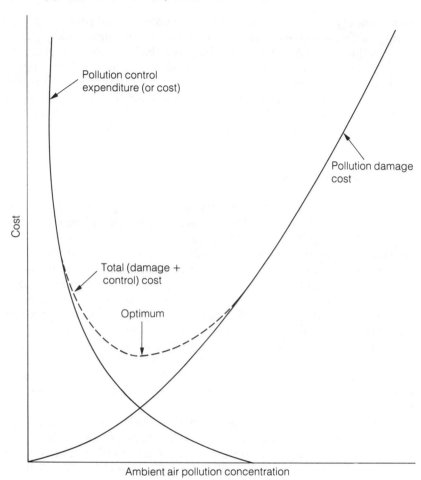

Figure 7.4 Schematic representation of the relationship between pollution-control expenditure, pollution-damage costs and ambient air-pollution concentration

7.6 Strategy Adoption

Although some attempts at tackling air-pollution problems date back to the last century or even earlier, most countries introduced their strategy for air-pollution control in the 1950s, 1960s or 1970s. Nearly all have adopted either the air-quality management strategy or the emission-standards strategy, or a combination of both (Martin, 1975; Campbell and Heath, 1977). At one extreme lies the Soviet Union which has over one hundred national air quality standards and few emission standards.

The United States is another strong advocate of the air-quality management approach, although it does have a large number of emission standards and has also recently introduced aspects of an economics-based strategy. In contrast, the United Kingdom has traditionally employed the emission-standards strategy with the concept of 'best practicable means' being introduced in the Alkali and Works Regulation Act of 1863. However, its entry into the European Community, which adopted the air-quality management strategy as its common policy in 1980, caused a radical change in the British approach to air-pollution control as it did also for Denmark and Ireland. It is likely that in the future there will be greater international uniformity of strategy adoption as the problem of long-range transport of air pollutants receives increasing attention.

There are many arguments as to which strategy is best (de Nevers et al., 1977; Wall, 1976b) but increasingly the conclusion being reached is that it is possible to make use of the advantages of two or more strategies. Persson (1977), for example, argues that for single emission-sources and small industrial areas with good atmospheric dispersion, emission standards and requirements on stack heights based on best practicable means will guarantee an air quality better than the air-quality standard. A special nondegradation policy is not needed. In large urban industrial areas, the air-quality standard should be the determining factor for regulations and systems planning concerning industrial development, energy supply, and transportation. Furthermore, in times of economic constraints, the incorporation of aspects of an economics-based strategy may be beneficial. In general, a national strategy evolves and changes over time and ultimately, the detailed nature of the strategy adopted by a country is unique to that country. The following chapters consider the nature of air-pollution problems in selected countries or groups of countries, and the strategy and tactics adopted to tackle air-pollution problems in order to achieve an acceptable level of pollution.

8 Pollution Control in the United States of America

8.1 The Early Years: Pollution as a Local Nuisance

Prior to the 1940s, although communities experienced severe air pollution, control measures were largely limited to the passing of local legislation or to private litigation claiming air pollution to be a common-law nuisance. Concern for air pollution focused exclusively upon visible emissions, especially smoke. The first municipal legislation prohibiting emission of 'dense' smoke was enacted in 1881 by Chicago and Cincinnati (Stern, 1982). By 1912, of twenty eight cities in the United States with a population of over 200,000, twenty three had programmes for the abatement of 'dense', 'black' or 'grey' smoke. With the increasing growth and concentration of activities that accompanied urbanization, pollution nuisances occurred more frequently and affected larger numbers of people, yet pollution was essentially perceived to be a problem of dense smoke emission to be dealt with at the local level. It was not until the California law of 1947 that a state law tackled air pollution other than dense smoke, and not until 1952 that Oregon introduced the first comprehensive state air-pollution control legislation.

8.2 California 'Discovers' Air Pollution

From 1943 onwards, Los Angeles began to experience increasingly frequent episodes of a brownish, hazy, irritating and altogether mysterious new kind of air pollution that was more persistent than, and quite different from, the instances of smokey fog which had troubled major urban centres from at least the mid-1800s onwards (Krier and Ursin, 1977). The reduced visibility accompanying these episodes prompted the common use of the term 'smog' (figure 8.1). Given that the term was originally coined from the words 'smoke' and 'fog', it is really a misnomer for photochemical air-pollution episodes.

During the Los Angeles smogs many thousands of people experienced eye and nose irritation, and action was demanded to curb these mysterious

Figure 8.1   Photochemical smog remains a problem in Los Angeles despite the introduction of stringent pollution-control policies
Source: Popperfoto

episodes. Optimistically, the mayor announced at an August 1943 conference 'that there would be an entire elimination of smog within four months' (Krier and Ursin, 1977). A plant manufacturing synthetic rubber and producing heavy visible emissions was closed down, seemingly as a scapegoat, a quick succession of ordinances aimed at sulphur dioxide and smoke emissions were passed in southern California, and the California Air Pollution Control Act of 1947 was enacted. The 1947 Act permitted, not required, air-pollution control districts to be set up and it introduced a programme to control pollution but at the same time keep industry in business as usual. However, effective action would be limited while the cause of the smog remained unknown. The smogs continued undiminished as no control measures were directed towards motor vehicles, whose role in producing smog was believed (as based on the experience of air-pollution research undertaken in eastern cities) to be relatively minor compared with industrial and municipal sources of pollution.

The first signs of guilt being attached to the motor vehicle as the cause of the smog occurred during November 1949 when the Universities of California and Washington State football teams played a game at Berkeley, north California. Many thousands of those attending experienced intense eye irritation, the hallmark of Los Angeles smog, but only at Berkeley. The weather was similar to that associated with severe smog episodes in Los Angeles yet no unusual industrial operations had occurred in Berkeley.

The only unusual occurrence was the vast numbers of vehicles converging on the football stadium producing an engine-idling, stopping and starting, traffic jam. It could only be concluded that the cause of the intense eye irritation was in some way related to vehicle exhaust emissions. Following this incident, and from research studies, Professor A. J. Haagen-Smit suggested that the Los Angeles smog arose because certain pollutants, principally from motor vehicles, oil refineries and backyard incinerators, were converted to eye- and nose-irritating pollutants by photochemical reactions. The important precursors were volatile hydrocarbons and oxides of nitrogen which, in the presence of sunlight, produced the irritating oxidant pollutants of ozone and PAN, among others. These findings were greeted with scepticism by the motor vehicle industry, the oil industry, and the vehicle-loving public. Nevertheless, by the mid-1950s, independent research confirmed Haagen-Smit's findings and ranked vehicle exhaust emission as the principal contributor to Los Angeles smog, even though the motor vehicle industry issued a series of findings between 1954 and 1959 which added confusion to the debate and caused a delay in the introduction of vehicle control measures.

Following Haagen-Smit's findings in 1950, the Los Angeles Air Pollution Control District was faced with a dilemma in its attempts to introduce an immediate and effective control programme. It could not require installation of control devices on vehicles sold in the county until a satisfactory device was available – and no such device was available. This meant that the smog problem could only be solved in the short term if the Los Angeles County restricted motor vehicle use. Consideration was given to measures such as encouraging car pools, creating a rapid transit system, prohibiting vehicles at certain times or in certain areas, and imposing a smog tax on vehicles. Reluctance to adopt any of these measures with any strong conviction highlighted the widespread belief in the Los Angeles community that the problem was a scientific, technological or engineering problem rather than a social, political or legal one. If solutions were to be found they would be of a technological nature.

Many such 'solutions' were put forward during the 1950s. Suggestions included removing the subsidence temperature inversion which trapped the smog, using ground-based fans, helicopters or thermal means; eliminating the sunlight and thus the smog by producing a gigantic parasol of white smoke laid by aircraft high over the city; removing the smog through tunnels in the mountains around the Los Angeles basin using huge fans; and seeding the air with some sort of agent that would 'neutralize' the smog. Although some of these technological fixes indicate a degree of desperation and hopelessness associated with the smog problem, the magnitude of the problem still existing in the Los Angeles basin in the 1980s prompted Heicklin (1981) to suggest the technological fix of introducing diethylhydroxylamine as a 'relatively inexpensive and safe' chemical inhibitor of smog.

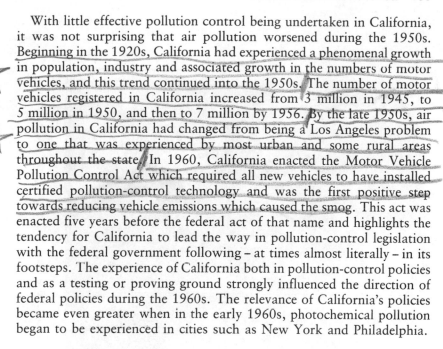

With little effective pollution control being undertaken in California, it was not surprising that air pollution worsened during the 1950s. Beginning in the 1920s, California had experienced a phenomenal growth in population, industry and associated growth in the numbers of motor vehicles, and this trend continued into the 1950s. The number of motor vehicles registered in California increased from 3 million in 1945, to 5 million in 1950, and then to 7 million by 1956. By the late 1950s, air pollution in California had changed from being a Los Angeles problem to one that was experienced by most urban and some rural areas throughout the state. In 1960, California enacted the Motor Vehicle Pollution Control Act which required all new vehicles to have installed certified pollution-control technology and was the first positive step towards reducing vehicle emissions which caused the smog. This act was enacted five years before the federal act of that name and highlights the tendency for California to lead the way in pollution-control legislation with the federal government following – at times almost literally – in its footsteps. The experience of California both in pollution-control policies and as a testing or proving ground strongly influenced the direction of federal policies during the 1960s. The relevance of California's policies became even greater when in the early 1960s, photochemical pollution began to be experienced in cities such as New York and Philadelphia.

8.3   The Federal Government Becomes Involved

Although the eye irritating smogs of Los Angeles in the 1940s received nationwide media coverage, it was not until a 1948 six-day smog in Donora, Pennsylvania, causing 6000 cases of illness and twenty deaths, that air pollution succeeded in gaining federal attention (Schrenk et al., 1949). The Donora episode, together with smogs in London in 1952 (causing 4700 deaths) and New York City in 1953 (resulting in 200 deaths – Greenburg et al., 1962), stimulated plans for a national air-pollution conference and the introduction of a few bills in Congress (none of them passed). The bills had merely called for further federal studies: control was considered a state and local concern. The Air Pollution Act which was finally enacted in 1955 reflected the same view: federal legislation left regulatory control efforts to the states and local government. It authorized $5 million annually for five years (later extended to 1964) to support research, provide technical assistance to public and private organizations, and provide for the training of technical personnel. Although this first national legislation placed the federal role as a passive supporter of state and local government policies, it marked the beginning of federal involvement which was to be transformed into an interventionist and policy-initiating role.

Figure 8.2 The four-day smog in New York City in November 1966 led to a state of emergency being declared
Source: Associated Press

By 1960, recognition that the pollution problem associated with motor vehicles had reached nationwide significance was highlighted by the enactment of the Motor Vehicle Exhaust Study Act (the Schrenk Act). This 1960 Act required the Surgeon General to undertake studies of the health effects of motor vehicle exhaust emissions. Air-pollution problems gained further national attention in December 1962 with the Second National Conference on Air Pollution. This meeting coincided with a London smog which caused 700 deaths and so was cited as evidence of the urgency with which air-pollution problems should be tackled. The Clean Air Act of 1963, the namesake for the major federal incursions to follow in 1970, expanded the research and technical assistance programme initiated in 1955, provided limited federal powers to investigate and abate pollution emissions endangering health and welfare, instructed the US Department of Health, Education and Welfare to develop criteria on the effects of air pollution and its control (advisory guidelines to be used or ignored by state and local government), and increased its research and development of devices and fuels to aid in the prevention of industrial and vehicle pollutant discharges. Federal attention to motor vehicles increased thereafter, with the motor vehicle being repeatedly charged with half of the blame for the nation's air-pollution problems.

The Motor Vehicle Pollution Control Act of 1965 brought the federal government into the business of controlling motor vehicle emissions, with national emission standards for new vehicles being set, starting with the 1968 model year. No state except California was allowed to adopt new vehicle standards more stringent than the federal ones: California's pollution problems were recognized as exceptional. The emission standards set for new vehicles gave due consideration to the technological feasibility and economic cost of compliance by the vehicle manufacturers and the standards which emerged were similar to those already adopted by California.

Pressure for stronger federal control of air pollution escalated throughout the 1960s. Pollution episodes continued unabated with, for example, a four-day episode in New York City in November 1966 causing eighty deaths and resulting in a state of emergency being declared by Governor Rockefeller (figure 8.2). By 1966 little more than half of the country's urban population enjoyed local air-pollution control; of almost 600 counties with a population greater than 50,000, less than ninety had control programmes, and most of these programmes were far from adequate (Krier and Ursin, 1977). As the Clean Air Act had left much to state initiative, congressional dissatisfaction with the rate of progress in some states led to the Air Quality Act of 1967 which at last required states to establish air-quality standards consistent with federal criteria, and then to devise implementation plans setting out ways of achieving the air-quality standards. If states failed to set or enforce standards, the

Department of Health, Education and Welfare could intervene. However, although federal government could now intervene, albeit only after pursuing lengthy procedural steps, the Act still left the primary responsibility for air pollution control to the states and local government.

Some states were not slow in following up the 1967 Act. California, which was exempt from the federal ruling that no state could adopt emission standards for new vehicles more stringent than the federal ones, passed the Pure Air Act of 1968, which contained explicit hydrocarbon, carbon monoxide and oxides of nitrogen emission-standards for post-1970 new vehicles. This state Act recognized the flaws of its earlier attempts at photochemical-pollution control which had not included an emission standard for oxides of nitrogen. Only exhaust emission standards for hydrocarbons and carbon monoxide had earlier been set and this ignored the fact that if hydrocarbons and carbon monoxide were controlled, emissions of oxides of nitrogen increased.

Although emissions of carbon monoxide, hydrocarbons and oxides of nitrogen can be reduced to a certain extent by modifications to the engine combustion, meeting the emission standards requires the fitting of catalytic converters. The simplest type, called an oxidation catalyst, is a canister fitted with a porous ceramic element which is coated with a thin layer of platinum, palladium or other noble metal of the platinum group. The exhaust gas is mixed with a little air from a small pump at a point just before it enters the canister. The platinum catalyzes the reaction of carbon monoxide and hydrocarbons with air to give carbon dioxide and water vapour. This type of catalyst does not affect oxides of nitrogen. To reduce oxides of nitrogen, a three-way catalyst is required. Its construction is similar to the oxidation catalyst except that it uses a different mixture of platinum metals, has no air pump, and requires a special mixture strength. The engine has to be electronically controlled to give a fixed level of oxygen in its exhaust. This is detected with a special sensor in the exhaust manifold. With the exhaust gas thus controlled the catalyst causes a reaction between oxygen and oxides of nitrogen on the one hand and carbon monoxide and hydrocarbons on the other. Oxides of nitrogen are reduced to harmless nitrogen, and the carbon monoxide and hydrocarbons are oxidized to carbon dioxide and water, as before (UK Royal Commission on Environmental Pollution, 1983).

By the end of the decade, despite progress by some states, advancement towards an effective nationwide pollution-control programme was slow. Ralph Nader's Study Group on Air Pollution estimated that the implementation phase would not be completed until well into the 1980s and the Air Quality Act was criticized for providing the Department of Health, Education and Welfare with only very limited powers to press the states into faster action. With growing public concern over air quality in cities and with prospects for only a slow improvement in the coming decade, the federal government was put under increasing public pressure to introduce a more effective pollution-control policy. Such a policy would be implemented only if the opposition of the very well organized and wealthy political lobbies of industry could be overcome. In contrast with earlier times, the Environmental Movement, peaking in 1969 and 1970, provided the political muscle necessary to counter the lobbying by industrialists.

8.4   The Environmental Movement of the Late 1960s

Prior to the mid-1960s, pollution issues received only limited and sporadic attention from a public whose interest was largely restricted to rather isolated localized pollution issues and the occasional air-pollution episode. However, during the late 1960s a dramatic and unprecedented increase in public concern for the environment occurred. Public opinion polls revealed the problem of pollution to have risen from ninth of the ten most serious problems facing the nation in 1965, to second in 1970 (Jones, 1973). Environmental concern sprang suddenly from nowhere to reach almost crisis proportions, not because of a sudden deterioration in the environment but because popular awareness escalated (Bowman, 1975). Sellers and Jones (1973) point out that 'the deterioration of the environment was discovered by the news media sometime in 1969' (figure 1.2). With the media thriving on doomsday and eco-crisis stories with the recurrent theme of 'immediate danger, immediate action' (Chase, 1973), public interest, and demands for prompt action, inevitably resulted: the 'Environmental' or 'Ecology' Movement was born.

Whether it was the media, leaders of interest groups, or trigger events (such as the sequence of New York City smogs during the 1960s, publication of numerous books which continued the theme first introduced in Rachel Carson's *Silent Spring*, or Ralph Nader's environmental reports) which generated the Environmental Movement, the media sustained public interest for environmental issues (figure 8.3). Such heightened public concern for the environment in 1969 ensured the emergence of the environment as a political issue. Environmental pressure-groups, aided by the grassroots backing of the general public and the critical monitoring of policy development by the media, were able to exert considerable pressure on government policy. Their impact on environmental policy development at the height of the Environmental Movement in 1969–70 was dramatic and resulted in the enactment of the innovative National Environmental Policy Act of 1969 and the sweeping Clean Air Amendments of 1970. Although the latter legislation is correctly termed the Clean Air Amendments, it is commonly referred to as the Clean Air Act, and this practice is followed here.

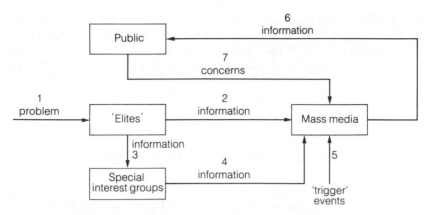

Figure 8.3   The elite-mass media-public interaction model
Source: Parlour and Schatzow, 1978

Enactment of strong legislation is only part of the process towards achieving an environmental clean-up; it also has to be applied effectively in following years. The Environmental Movement was too short-lived to ensure the rigid implementation of the 1970 Clean Air Act and this is one reason why the Clean Air Act was subsequently relaxed and weakened during the 1970s. By 1974, the 'national' problem of pollution was overtaken by the problems of 'integrity in government' (triggered by the Watergate scandal), 'energy shortage', 'crime', and 'taxes and spending' (Sandbach, 1980). Even by 1972, energy and economic issues displaced environmental issues as more newsworthy, being deemed by the media to be of greater and more immediate public interest (Parlour, 1980). The media no longer played the role of sustaining public concern for environmental issues. As Gregory (1972) points out, 'the media thrive on fashion, and do much to create it. Fashions are ephemeral; new ones regularly drive out the old'. Moreover, public demands for federal action were fulfilled with the enactment of major environmental legislation. Pollution problems may not have been solved yet but many of the public felt satisfied that they would be, now that legislation existed.

The sudden rise and sharp decline in public concern for pollution seems to follow what happens to most major social problems as exemplified by the 'issue attention cycle' (Downs, 1972; O'Riordan, 1976) or the 'worry bead hypothesis' (Kates, 1977). Nevertheless, although the state of the environment is no longer viewed as the crisis issue that it was in 1969–70, public support for environmental protection remains considerably stronger that it was prior to the emergence of the Environmental Movement (Anthony, 1982).

8.5   Pollution as a Political Issue

The development of federal air-pollution policy prior to the 1970s was characterized by incrementalism. Following some reluctance to become involved in air-pollution control, the federal government opted for a research role in 1955 and only slowly did it begin to intervene with the control policies adopted by the states. At first it left the initiative to state and local government but then it progressively took away more and more of it. Each step was a careful reaction to circumstances, whether these were pollution episodes or new research findings as to the causes of air-pollution problems. Policy development was incremental in as much as it adopted the smallest, least disruptive step with the least apparent cost (Lundqvist, 1980). Policy, it was believed, had to be based on sound scientific evidence; otherwise, it would not be 'fair' to industry. All this changed in 1969 and 1970 with the emergence of the Environmental Movement which made the environment a political or 'vote-catching' issue, and which produced massive public pressure for fast, dramatic action, the nature of which is usually restricted to crisis situations (Gunningham, 1974; Jones, 1973). Environmental policy subsequently underwent dramatic shifts in both stride and direction.

The upsurge in public concern over environmental-quality issues during the latter half of 1969 led to the enactment of the National Environmental Policy Act, which was passed with limited debate or opposition despite its being so fundamental and innovative a measure. The act attempted to 'create and maintain conditions under which man and nature can exist in productive harmony, and fulfil the social, economic and other requirements of present and future generations of Americans (Goodenough, 1983). It required all federal agencies to take into account the environmental impact of all their proposed actions by preparing Environmental Impact Statements which had to be approved before the construction of 'major' projects. Although this requirement was directly applicable only to federal agencies or projects requiring federal licences, in practice it has been extended to all major schemes of industrial expansion and new construction. Similarly indicative of the public interest with the environment was the passing of the Environmental Education Act of 1970 which provided means and funds for the development of new and improved curricula for an understanding of environmental-quality problems.

By 1970, environmental protection was being proclaimed by the press, radio and television as 'the issue of the day'. The 'environment' had become a major political issue, and with the media constantly monitoring the progress of policy development, this resulted in active competition between elected officials to produce and be credited with strong legislation. President Nixon led the way with a timely 'message on environmental quality' for his February 1970 State of the Union Address. Nixon summed up public opinion: 'Air is our most vital resource, and its pollution is our most serious environmental problem. Existing technology is less advanced . . . but there is a great deal we can do within the limits of existing technology – and more we can do to spur technological advance.' The resulting radical pollution-control programme outlined by Nixon, calling for a 90 per cent reduction in vehicle emissions by 1980, not only led to him being credited (albeit briefly) as policy-initiator of an environmental clean-up but also provided him with the chance to deal a blow to one of his most important opponents in the 1972 elections, Edmund Muskie.

Much of Muskie's political fame was earned from his leadership in environmental affairs, which was now being eroded not only by Nixon's proposals but also by a Ralph Nader report on air pollution which strongly criticized Muskie's role in the past and just stopped short of accusing him of selling out to industrial polluters (Lundqvist, 1980). Under this considerable pressure and criticism Muskie seemed to be left with no political option but to enter the game of policy escalation. He modified the Nixon bill (subsequently to become the Clean Air Act of 1970) to require 90 per cent reduction in vehicle emissions below 1970 federal standards for hydrocarbons and carbon monoxide by 1975, with a similar reduction in oxides of nitrogen below 1971 emission levels a year later. He justified the very strict timetable (suspensions were possible for one year only) on the basis of the need to place the protection of public health above considerations of technological and economic feasibility. The Nixon Administration, now on the defensive, asked for the Senate deadlines to be relaxed, but with strong public and media backing, the policy presented by Muskie was enacted. Claims that the bill placed unreasonable demands on the motor vehicle industry in seemingly disregarding the costs of compliance, at a time when the industry was already facing intense foreign competition, were dismissed. As Senator Griffin claimed:

> I think one of the problems with this legislation right now is that too many of the discussions with regard to the bill are being made on a political basis. When everybody is for clean air and against pollution [it is indeed] difficult politically to vote for any amendment that would be characterized by the press as weakening the bill.

8.6   The Clean Air Act of 1970

The Clean Air Act of 1970, swept into enactment by the political strength of the Environmental Movement, marked the beginning of a new – and the present – era of pollution-control policy. A dramatic expansion of federal intervention took place, and in order to cope with the new

Table 8.1   National Ambient Air Quality Standards (NAAQSs) in the United States

Pollutant[f]	Period of measurement	Primary standard (health-related)		Secondary standard (welfare-related)	
		$\mu g/m^3$	ppm	$\mu g/m^3$	ppm
Carbon monoxide[a,b]	8 hours	10,000	9	None	None
	1 hour	40,000	35	None	None
Nitrogen dioxide[c]	year (arith. mean)	100	0.053	Same	Same
Ozone[d]	1 hour	235	0.12	Same	Same
Sulphur dioxide[a]	year (arith. mean)	80	0.03	None	None
	24 hours	365	0.14	None	None
	3 hours	None	None	1300	0.50
Total suspended	year(geom. mean)	75	-	60	-
particulate[a] (TSP)	24 hours	260	-	150	-
Lead[e]	3 months	1.5	-	Same	Same

Standards for periods of 24 hours or less may not be exceeded more than once per year, except that ozone may use a three-year statistical average to determine if exceeded.

[a]NAAQS set on 30 April 1971.
[b]The EPA has proposed a revision which would reduce the 1-h standard from 35 to 25 ppm (29,000 $\mu g/m^3$) and increase the number of allowable exceedances of the 9ppm 8-h standard from one to five.
[c]Oxides of nitrogen are measured as nitrogen dioxide.
[d]NAAQS for photochemical oxidants (measured as ozone) set on 30 April 1971 at 0.08 ppm but revised in February 1979 and applicable only to ozone, the major and most easily measured oxidant.
[e]NAAQS set on 5 October 1978.
[f]A primary NAAQS for hydrocarbons (non-methane) of 160 $\mu g/m^3$ or 0.24 ppm for 3 hours not to be exceeded more than once per year was set in April 1971 but was rescinded in January 1983.
Source: after US EPA, 1986

legislation, a special pollution-control agency, the Environmental Protection Agency (EPA), was established in December 1970. The major points of the Act included six main features.

(1) Uniform National Ambient Air Quality Standards (NAAQSs) were to be set by the EPA to protect public health and welfare (table 8.1). Whereas the 1967 Act allowed states to devise their own standards, taking into account federal advice, the 1970 Act empowered federal government to set uniform standards to apply to all states. This overcame the concern that some states might compete for industry by setting lax requirements, so becoming 'polluter havens', or that large industry in some states might lobby for the establishment of permissive standards. Each state was required to submit a State Implementation Plan (SIP) indicating how its control programme would bring about the attainment of the NAAQSs by 1975. If a state failed to develop or execute a satisfactory plan, the EPA had the authority to intervene directly.

(2) Uniform national standards of performance for new industry were to be established. Uniformity avoided the unfair economic competition which might have resulted if the setting of standards had been left to the

states. Nevertheless, the Act allowed states to set more stringent standards if they wished to do so. Federal performance standards were applied to new stationary sources or to existing plants undertaking modifications. These New Source Performance Standards (NSPS) have been promulgated and revised regularly by the EPA since 1971. Before a new stationary source could begin operation, state or federal inspectors were required to certify that the pollution controls would function adequately.

(3) National emission-standards for hazardous air pollutants were to be established by the EPA and would apply to existing as well as to new industrial plants. Arsenic, asbestos, benzene, beryllium, mercury, radionuclides, and vinyl chloride have all been designated as hazardous according to section 112 of the Act, although numerous pollutants have been listed as potentially hazardous and await the setting of standards by the EPA. The EPA has received considerable criticism concerning the length of time it takes to set such emission standards.

(4) Uniform restrictions on emissions from new motor vehicles were to be set. The stringency of these restrictions reflected public opinion that the motor vehicle was the primary villain in air-pollution problems, accounting for between 50 and 60 per cent of total US air-pollution emissions. The emission standards were set without regard to the constraints of technological or economic feasibility that had previously influenced policy-making. Nevertheless, $89.1 million was authorized for a research programme (1970–5) to develop a low-emission alternative to the current internal combustion engine (Wark and Warner, 1981).

(5) Citizens were permitted to take legal action against any person, including the United States government (for example, the EPA), alleged to be in violation of either emission standards or an order by the administrator. Environmental pressure-groups, such as the Sierra Club, have used this right on several occasions to press the EPA to take action: for example, environmental groups argued through the courts that Congress intended the EPA regularly to issue air-quality standards for pollutants that were judged to be widespread and hazardous to public health. Litigation between 1974 and 1978 concerning the need for an ambient air-quality standard for lead resulted in the EPA being ordered to set a NAAQS for lead (table 8.1).

(6) A $30 million research programme was initiated to assess the causes and effects of noise pollution on public health and welfare. The subsequent research findings presented to Congress resulted in the Noise Control Act of 1972 (later expanded into the Quiet Communities Act of 1978) which required the EPA to set standards of tolerable noise levels for all types of new equipment and machinery. The initial objective of the noise-control programme was to reduce environmental noise everywhere to below 75 db(A), a level at which there is a risk of hearing damage (Cherfas, 1980).

8.7   Problems Develop with the State Implementation Plans

The 1970 Act required the EPA to prescribe NAAQSs which were not to be exceeded in any region more than one day per year, nor during more than a limited period of time within that day (table 8.1). Primary or public health standards were aimed at protecting the most susceptible part of the population from adverse effects, with an adequate margin of safety being included, while the more stringent secondary standards were aimed at protecting public welfare, such as damage to vegetation, wildlife, materials, and so on. In 1971, the EPA promulgated NAAQSs for six 'criteria' pollutants and the strict legislative timetable required states to submit implementation plans which would achieve primary standards for each pollutant by 1975, or, if the deadline was extended as the EPA was authorized to do so, by 1977. Once primary standards were attained, the state was expected to attain the secondary standards 'within a reasonable time'. In contrast with the tech-nological fixation which had previously dominated the control strategies of many states, the EPA stressed that the State Implementation Plan (SIP) should consider incorporating transportation controls to reduce the distance travelled by all vehicles, new and old alike. Suggestions included the promotion of mass transit systems, parking restrictions, staggered working hours, commuter taxes, and even petrol rationing. Land-use controls should also be examined. Economic incentives or disincentives such as pollutant emission charges or taxes should receive more attention than they had received previously. During emergency periods or times of air-pollution episode potential, consideration should be given to temporarily closing polluting plants or requiring fuel-switching to be adopted by larger industry.

Having devised the SIP, the state had to present monitoring and modelling data indicating that its control programme would bring about the attainment of the primary NAAQSs. The models employed to develop the control programme vary from the simple 'proportional' or 'rollback' model which assumes, for example, that a region with sulphur dioxide levels twice the NAAQS will attain the standard if total sulphur dioxide emission in the region is halved, to more complex diffusion or dispersion models. Dispersion models predict ambient concentrations of pollutants from emission inventories of pollution sources, meteorological conditions, and topographical considerations (refer to section 7.2).

Given the strict timetable for submission of the SIP and attainment of NAAQSs, it was not surprising that some states would face enormous difficulties, not least the state of California. The seemingly impossible 1975 (or 1977 if extended as allowed under the act) goal for attainment of the photochemical-oxidant standard is highlighted by data for Los Angeles. In 1970 the Los Angeles County exceeded the California standard for oxidant (0.10 ppm) on 241 days of the year. Under the 1970 legislation, the county was required by 1975 (or 1977) to exceed the more demanding federal standard (0.08 ppm) no more than one day per year! Los Angeles Air Pollution Control District believed that the state programme could, with no interference, achieve marked improvement in the level of photochemical smog by 1980, and that by 1990 the atmosphere would meet the ambient air-quality standard (Krier and Ursin, 1977). Others were less optimistic, pointing out that population and industrial growth would offset the improvements being made. California's SIP was rejected by the EPA because it did not provide for attainment of the photochemical-oxidant standard for Los Angeles. Following legal suits by citizens' organizations, the EPA reluctantly set out to devise an alternative plan. It increasingly realized that no technical measures could assure compliance by the extended 1977 deadline, and that since an adequate mass transit system could not be developed in time, only petrol rationing would lead to attainment. Petrol rationing of over 80 per cent during the smog season from May to October was suggested! Seemingly valid claims by the Los Angeles County that such a proposal was economically and politically unrealistic contributed to undermining the EPA's demands for maintaining strict attainment deadlines.

Progress by states towards developing an acceptable SIP was further hampered by the Arab oil embargo of 1973. Several strategies were available to states to bring stationary sources into compliance with air-quality standards. Land-use planning may be used to regulate the number and size of polluting sources within any given area, low-sulphur fuels may be employed, continuous control equipment may be installed (for example, flue gas desulphurization systems), or intermittent control systems may be used, such as switching to low-sulphur fuels during unfavourable meteorological conditions. Given these choices, many states opted to control sulphur dioxide emissions by regulating the maximum sulphur content of the fuel allowed to be burned. However, domestic supplies of low-sulphur oil were inadequate to meet the potential demand and the oil embargo of 1973 dramatically highlighted this problem. The increasing price of low-sulphur oil and the reluctance to rely upon imported oil in the future brought a growing demand by industry and by some politicians for the adjustment of the 1970 Act requirements in the light of economic and energy realities as well as technological practicability (Lundqvist, 1980). Subsequently, the Energy Supply and Environmental Co-ordination Act of 1974 allowed short-term suspensions of air-pollution standards through June 1975, if 'clean' fuels were not available.

8.8   Motor Vehicle Emission-standards: A Change from the Desirable to the Feasible

During and following the enactment of the 1970 Act, the motor vehicle industry launched a vigorous campaign to publicize its claim that the attainment of the vehicle emission-standards for carbon monoxide and hydrocarbons by 1975, and for oxides of nitrogen by 1976, was impossible. Whereas Muskie was convinced that industry could comply with this 'technology-forcing challenge' (had not industry developed the technology to place a man on the moon following the challenge by President Kennedy?) – and the strong anti-pollution public opinion of 1970 added weight to this claim – counterclaims gained increasing attention during the 1970s. As public concern for the environmental crisis gave way to the energy crisis (the oil embargo of 1973) and escalating unemployment, claims that the stringent vehicle emission requirements were causing economic hardship in the motor vehicle industry and affecting its competition against foreign producers were given a more sympathetic hearing.

In April 1973, the EPA recognized the pressure for relaxation of the deadlines by granting a one-year extension for carbon monoxide and hydrocarbons. At the height of the energy crisis, the enactment of the Energy Supply and Environmental Co-ordination Act of June 1974 suspended the emission standards until 1977 and 1978 to allow vehicle manufacturers to devote more time to improving fuel economy. In March 1975, a brief concern (shown to be unfounded one month later) that the use of oxidizing catalysts on vehicles was encouraging the production of sulphates which could pose a considerable threat to public health led the EPA to suspend the carbon monoxide and hydrocarbon emission-standard deadlines for one year. Economic-practicability and technological-feasibility arguments gained increasing public, media and federal attention as the energy crisis was followed by the economic recession such that the Clean Air Amendments were passed in August 1977. These amendments postponed the original 1975–6 carbon monoxide and hydrocarbon standards until 1980–1, and established a relaxed standard for oxides of nitrogen (the revision readjusted the balance between air quality and fuel economy by raising the emission limit from 0.4 to 1.0 grams per mile) to take effect in 1981 or even later (Table 8.2).

8.9 The Clean Air Amendments of 1977

Increasing pressure for the relaxation of the strict deadlines for motor vehicle emission-standards and for the attainment of NAAQSs resulted in the Clean Air Amendments of 1977. Vehicle emission-standards for carbon monoxide and hydrocarbons were suspended until 1980–1 and the original oxides of nitrogen standard became merely a long-term research objective. With most areas of the country in 1977 not having attained the NAAQS for at least one pollutant, extensions to the attainment deadlines were inevitable. Non-attainment areas were given until July 1979

Table 8.2   Exhaust emission-standards for vehicles

Model year	Standard (gm/mile)[a]			Percentage reduction from pre-control vehicle		
	HC	CO	NOx	HC	CO	NOx
Pre-1968[b]	8.2	88.5	3.4	–	–	–
1968-71	4.1	34.0	None	50	62	0
1972-4	3.0	28.0	3.1	63	69	9
1975-6	1.5	15.0	3.1	82	83	9
1977-9	1.5	15.0	2.0	82	83	41
1980	0.41	7.0	2.0	96	92	41
1981	0.41	3.4[c]	1.0[c]	96	96	76

[a]All standards are based on current test procedures.
[b]Average in-use emission levels of gasoline vehicles.
[c]EPA allowed to waive these standards.
Source: US National Commission on Air Quality, 1981

to have an approved SIP which provided for attainment of the primary NAAQSs, by December 1982. If, despite the implementation of all reasonably available measures, including the introduction of a motor vehicle inspection and maintenance programme, a state could not attain primary standards for carbon monoxide and photochemical oxidants by 1982, it had to submit a second SIP which provided for attainment by December 1987. Despite being a non-attainment or 'dirty' area, new sources of pollution could be built providing that first, the proposed plant installed pollution-control technology which ensured the 'lowest achievable emission rate' (LAER); and second, that the proposed emissions were offset by reductions in emissions from existing sources in the area (the 'emission offset' policy).

Although attention was naturally directed towards improving the 'dirty' area, the Sierra Club successfully pressed through the courts that the EPA should ensure that 'clean' or attainment areas should not suffer further degradation (Stern, 1977). Congress incorporated this principle into the 1977 Amendments in the form of the Prevention of Significant Deterioration (PSD) programme. This assures a level of protection for clean areas beyond that provided by the primary and secondary ambient air-quality standards (Ward, 1981). Not only does the programme require that new sources of pollution must use the 'best available control technology' (BACT) to minimize their impact on air quality, but it also limits the additional pollution which can be added to the air. The PSD programme specifies three classes of areas:

> Class I   Areas where almost no change from current air quality will be allowed. 'Pristine' areas include National Parks and National Wilderness Areas and cover approximately 1 per cent of the federal land area.

Class II   Areas where moderate change or moderate economic growth will be allowed, and where no stringent air-quality constraints are desirable.

Class III   Areas where substantial industrial growth will be allowed and where the increase in concentrations of pollutants up to federal standards will be insignificant. As all areas were initially classified as Class I and II, the EPA left it to the states to redesignate some Class II areas as Class III, or even to upgrade such areas to Class I (Stern, 1977).

Additional air pollution in each of these classes must not exceed specific air-quality 'increments', thereby assuring a level of air quality which should in all cases be cleaner than that required by the NAAQSs. For sulphur dioxide and suspended particulates, the allowable increments equal approximately 2–5 per cent, 25 per cent, and 50 per cent, respectively, of the NAAQSs (table 8.3).

Table 8.3   NAAQSs and allowable PSD increments[a] ($\mu g/m^3$)

Pollutant	Controlling or determinative NAAQS	Class I	Class II	Class III
Sulphur dioxide				
Annual	80	2	20	40
24-hour	365	5	91	182
3-hour	1300	25	512	700
Total suspended Particulate matter				
Annual	75	5	19	37
24-hour	150	10	37	75

[a]Increments averaged for 24-hour and 3-hour periods may not be exceeded more than once a year. These short-term standards are generally the most difficult for industrial sources to meet.
Source: Ward, 1981

Both environmentalists and industrialists agree that the PSD programme is complex. Industry regards the PSD programme as 'the most burdensome requirement of the Clean Air Act' (Hart, 1982). To comply with the provisions of the programme, a company must first spend a year monitoring air quality in the proposed area and employ elaborate computer models to demonstrate that pollution from the proposed plant will stay within the allowable increments. Even so, environmentalists point out that 75 per cent of all PSD permits were issued in ten months or less.

Protection of air quality in pristine areas is also achieved through the use of visibility standards introduced by the EPA in 1984 for thirty three states and one territory (the Virgin Islands) in which National Parks and Wilderness Areas are deemed to need visibility protection as mandated by the 1977 Clean Air Amendments. States are required to revise SIP requirements as needed to make reasonable progress towards the visibility standards, and to develop long-term, (10–15 year) strategies for meeting these goals. The EPA adopted a phased approach with regard to visibility protection. Initial action has focused on 'plume blight'. Plume blight is defined as an identifiable coherent plume which is objectionable in an area in which visibility is an important value, because of its colouration and intrusive nature. It is usually caused by the emission of particulates or nitrogen dioxide. It is controlled by requiring the offending pollution source to employ the 'best available retrofit technology' (BART). A much more complex and far more serious cause of long-range visibility impairment is that of 'regional haze' caused by sulphate and nitrate particles. This problem is an integral part of the acid rain problem and will only be solved when emissions of sulphur dioxide and oxides of nitrogen are substantially reduced.

8.10   Pollution-control Policy: Job-taker or Job-maker?

One of the criticisms frequently made by industry concerning pollution-control policy is that it causes severe economic dislocation and unemployment. A car sticker slogan sums this up simply: 'Are you poor, hungry, out of work: eat an environmentalist'. Industrialists argue that the expensive pollution-control equipment which companies have to purchase, install and maintain leads to plant closures (Stafford, 1977). Controls required on new stationary sources delay or even prevent the construction of major industrial facilities, such as power stations and oil refineries, thereby aggravating the energy situation and generally restricting the economy's ability to expand. In general, whereas previously a site for a new plant could have been selected in three to six months, it may now take two or more years. By that time it may be too late to get into the market place with the product. Companies faced with massive investment in new pollution-control equipment, if a plant is to be located in a new location, may well find it cheaper and simpler to expand pollution at existing facilities, thereby adding to industrial inertia (Elsom, 1983).

Pollution abatement costs for industry can indeed be immense. Most industries commit in excess of 5 per cent of their total capital expenditure, with some industries committing far more, such as the paper industry (43 per cent), non-ferrous metals (23 per cent), iron and steel (14 per cent) and chemicals (12 per cent) (Starkie, 1976). In 1985, capital expenditure for air-pollution control was estimated to have cost industry $10.5 billion, with operation and maintenance costs at about the same level. According to the EPA the nation will have to spend $256 billion (at 1981 prices) between 1981 and 1990 to meet federal air-quality standards. Such costs compare with the $18 billion a year saved in wages and productivity which would have been lost due to sickness caused by air pollution and possibly $51 billion for the annual benefits (reduced medical treatment; less damage to buildings, crops and forests) to the nation as a whole of a less polluted society (Costle, 1979).

Criticisms of the pollution-control regulations frequently imply that the high cost of compliance leads to plant closures and curtailments, which in turn translate into unemployment. However, in reality the number of jobs lost compared with those gained in the major growth industry of pollution control is small. Between January 1971 and June 1981, the EPA identified 153 closures or curtailments in firms of twenty five or more workers, totalling 32,611 workers having lost their jobs (Kazis and Grossman, 1982). Most of these reported lay-offs were in four industries: chemical, paper, primary metals, and food processing. However, environmental regulation was only one of the reasons for the closures, for many firms which closed were the older inefficient or obsolete plants (Harris, 1981). Moreover, as many as 40 per cent of the lay-offs were rehired by their original companies at other plants. The numbers involved, not ignoring the personal distress associated with unemployment, seems small compared with other causes of lay-offs. Thus, for example, the Reagan Administration's 1982 budget cuts alone led to one million people becoming unemployed in both private and public sectors. In contrast, it has been estimated that between 0.5 and 1.1 million people are employed in private and public pollution-control (Harris, 1981; Kazis and Grossman, 1982).

8.11   EPA Flexibility: The Emissions Trading Policy

The 1970 Act as enacted offered few compromises in its aim to give the nation a healthy atmosphere by the mid-1970s. Only considerable pressure during the energy and economic crises and the subsequent desire for energy self-sufficiency and economic recovery achieved any relaxation in the aim: the primary goal of a clean atmosphere remained largely the same but the stringent deadlines for achieving this were sacrificed. Once this trend for the relaxation of the 1970 legislation began, the EPA considered that the only way in which it could slow down or stop this erosion, short of a resurgence in the Environmental Movement, would be to show more flexibility in its administration of the Act, especially towards new industry. Flexibility would take the form described by McAfee (1982) in that 'clean air regulations should tell business what do do, not how to do it – what goals to reach, not how to reach them.'

The 'bubble' strategy introduced in 1979, and incorporated into the Emissions Trading Policy of 1982, exemplifies the flexible approach by giving plant managers considerable freedom in finding the cheapest, most efficient way of meeting pollution-control standards (Behr, 1979; Liroff, 1980). Normally, the EPA and state control agencies set limits on emissions from each vent, smokestack, or other source of pollution in a plant. The bubble strategy considers the plant, or a series of plants, to be enclosed in a bubble and the EPA sets limits on the total discharges of each type of regulated pollutant, leaving it up to the plant manager to decide how to reach the goals (figure 7.3). Trade-offs between sources of pollution within the bubble must involve the same pollutant such that plants are not allowed to offset, say, reduced sulphur dioxide emission against having to place expensive controls on toxic emissions.

By 1983, the EPA had already either approved or proposed to approve 150 bubbles, with a saving to industry of $500 million. It is even conceivable that in the future the bubble policy may be applied to an entire city. One of the successes of the bubble policy to date includes a plant in Providence, Rhode Island, where $2.7 million was saved by replacing expensive low-sulphur oil at its two plants, one with high-sulphur oil and one with natural gas, which also resulted in a net reduction of the sulphur dioxide emission (Smith, 1981a). Initially, the EPA barred the trade-off between smokestack or plant emissions and windblown dust, arguing that industrial processes emitted smaller particles that stay in the air longer and lodge more deeply in the lungs. The agency was forced to reverse its decision under legal pressure from the steel industry, largely because the NAAQSs for suspended particulates did not make a distinction between particulate sizes, even though smaller particles are known to be more harmful. This ruling allowed Armco (one of the biggest steel producers) to use unconventional ways of controlling its particulate levels at its plant in Middletown, Ohio. Instead of spending $7 million on back-up equipment to catch the particles that its existing scrubbing plant could not cope with, it spent $1.7 million by periodically spraying its outdoor piles of iron ore and coal, by paving or periodically wetting down the company roads, by planting trees and grass to trap windblown particulates, and by arranging shuttle buses so that employees reduced the use of their own vehicles around the site.

The Emissions Trading Policy of 1982 also integrates the Emission Offset Interpretation Ruling of 1976 (amended in 1979 to incorporate banking). The Offset policy recognized that air is a scarce resource and used market forces to accommodate growth without increasing total pollution (US National Clean Air Coalition, 1981). Sponsors of a proposed new plant were required to offset the pollution which would result from their plant by: (1) reducing emissions from its own plant in the area; or (2) paying another company to reduce its emissions in the area; or (3) purchasing an old plant in the area and simply closing it down.

An example of an offset agreement is that between General Motors and local oil companies in Oklahoma City, which enabled the car company to build a new $400 million assembly plant there. The oil companies agreed to reduce their hydrocarbon emissions sufficiently to allow the car plant to be built. Offsets create an unofficial market in pollution or emission-reduction credits. Many firms using the bubble policy have found that the resulting pollution was less than what the law allows, a circumstance that grants them a credit towards added pollution in the future. As credits proliferated, the EPA decided to establish a banking or brokerage network through local governments for selling the credits from one firm to another, with the result that Louisville, San Francisco and Seattle initiated banking programmes. The use of market principles succeeds in facilitating new growth in an area under stringent pollution limits and permits compensation – at whatever the market will pay – for voluntary efforts to reduce pollution.

The Emissions Trading Policy marks a new chapter in flexibility in administrating federal regulations. Further innovations being considered include the issuing of auctionable pollution permits. Thus, for example, rather than barring production above current levels of, say, asbestos and chlorofluorocarbons at each plant throughout the nation, a ceiling on national production could be imposed with the rights to manufacture within the total being auctioned off to the highest bidders. The resultant high prices would force many out of the market and ensure that manufacturing rights went to those firms to whom they were economically most important (Smith, 1981b). This version of a marketable discharge licence process employs a national airshed, but it could be applied at a state or local level and for other pollutants in the future.

8.12 Progress Towards Improving Air Quality

Approximately 10 per cent of counties in the United States had not attained one or more of the primary ambient air-quality standards by 1984. This means that tens of millions of people still breathe 'unhealthy' air (figure 8.4). Nevertheless, the national level of air quality has on average improved significantly since the Clean Air Act of 1970 (US EPA, 1986). The national annual average concentration of total suspended particulate (TSP) decreased from about 88 μg/m^3 in 1960 to 50 μg/m^3 in 1985. Between 1975 and 1984, for which data are more accurate, a decrease of 20 per cent was measured using 1344 sites (figure 8.5). During that period, estimated emissions of particulates decreased by 33 per cent (figure 8.6; table 8.4).

However, national trends of TSP do conceal that during the 1970s 'dirty air got cleaner and the clean air got a little dirtier' (Lave and Omenn, 1981),

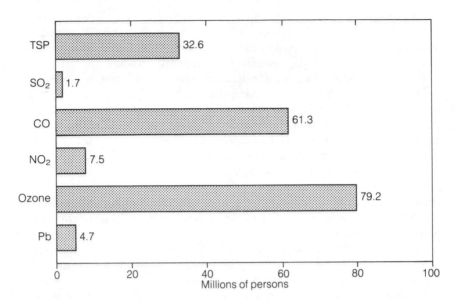

Figure 8.4 Numbers of persons living in counties with air-quality levels above the primary air-quality standards in 1984
Source: US EPA, 1986

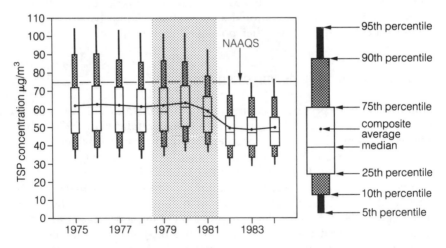

Figure 8.5 National boxplot trend in annual geometric mean TSP concentrations, 1975-84. The period 1979-81 is stippled to indicate that these data may be biased on the high side due to the use of glass fibre filter during monitoring in this period
Source: US EPA, 1986

Table 8.4 National annual emissions by pollutant and source in 1975 and 1984 (million tonnes/year)

Pollutant and source	1975	1984
Total suspended particulate		
Transportation	1.4	1.3
Fuel combustion	2.7	2.0
Industrial processes	5.0	2.5
Solid waste	0.6	0.3
Miscellaneous	0.7	0.9
Total	10.4	7.0
Sulphur oxides		
Transportation	0.6	0.9
Fuel combustion	20.3	17.3
Industrial processes	4.7	3.1
Solid waste	0.0	0.0
Miscellaneous	0.0	0.0
Total	25.6	21.4
Carbon monoxide		
Transportation	62.0	48.5
Fuel combustion	4.4	8.3
Industrial processes	6.9	4.9
Solid waste	3.1	1.9
Miscellaneous	4.8	6.3
Total	81.2	69.9
Oxides of nitrogen		
Transportation	8.9	8.7
Fuel combustion	9.4	10.1
Industrial processes	0.7	0.6
Solid waste	0.1	0.1
Miscellaneous	0.1	0.2
Total	19.2	19.7
Volatile organic compounds		
Transportation	10.3	7.2
Fuel combustion	1.0	2.6
Industrial processes	8.1	8.4
Non-industrial organic solvent use	1.9	1.8
Solid waste	0.9	0.6
Miscellaneous	0.6	0.9
Total	22.8	21.5
Lead		
Transportation	122.6	34.7
Fuel combustion	9.3	0.5
Industrial processes	10.3	2.3
Solid waste	4.8	2.6
Total	147.0	40.1

Source: US EPA, 1986

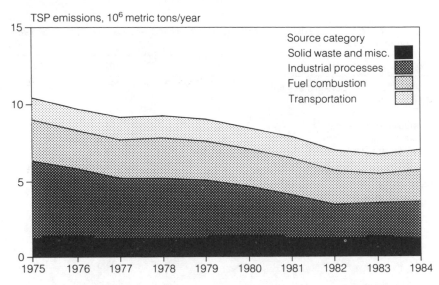

TSP emissions, 10^6 metric tons/year

Figure 8.6   National trend in particulate emissions, 1975–84
Source: US EPA, 1986

for it was not until the 1977 Amendments that the EPA had the authority
to prevent the deterioration of air quality in pristine areas. Furthermore,
the deterioration of visibility in some areas and the increased occurrence
of regional hazes in rural areas, resulting from the increase of sulphates
and nitrates, reflected the growth in emission sources, especially fossil-
fuel power stations, outside metropolitan areas. Although metropolitan
areas have in general shown considerable improvements since the 1970s,
the degree of success of pollution-control programmes has varied from
one area to another. Regional improvements in levels of TSP during the
1970s may in some instances be clearly related to the industrial recession:
for example, air quality in Beaver Valley Air Basin in Pennsylvania, which
is located northwest of Pittsburgh, shows a remarkable similarity to raw
steel production in the neighbouring Pittsburgh area from 1972–80 (figure
8.7). Not all regional changes in TSP concentrations are related to human
activities. The eruptions of Mount St Helens, for example, were responsible
for a 7 per cent increase in TSP in the EPA Region VII during 1980 (US
EPA, 1983b). Similarly, drought periods contribute windblown dust to
the particulate burden in some areas.

Average annual levels of sulphur dioxide at 229 urban sites showed
a reduction of 36 per cent for the period 1975–84 (figure 8.8),
while emissions decreased by 16 per cent over the same period (figure
8.9). Air-quality improvement is clearly related not only to regulatory
efforts which encouraged or required both power stations to install flue gas

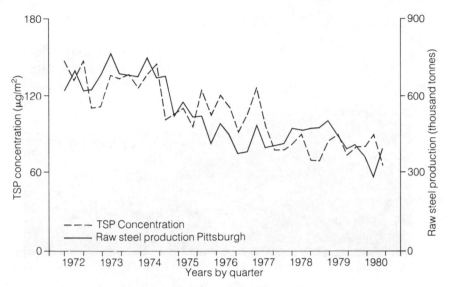

Figure 8.7 The relationship between total suspended particulate concentrations and raw steel production in Pittsburgh, 1972–80
Source: US EPA, 1983

desulphurization systems and industry to use fuels with a lower sulphur content, but also to the use of cleaner fuels in residential areas. Nationally, the urban sulphur dioxide problem has diminished to the point where only a small number of urban sites now exceed the air-quality standard. Remaining sulphur dioxide problem sites are found near large industrial

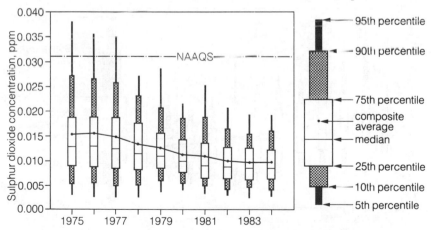

Figure 8.8 National boxplot trend in annual average sulphur dioxide concentrations, 1975–84
Source: US EPA, 1986

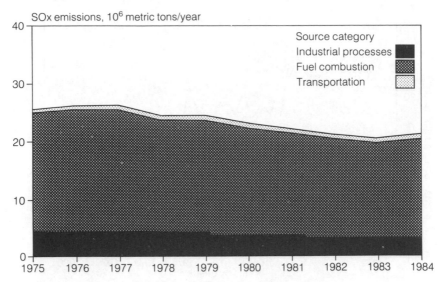

Figure 8.9 National trend in sulphur dioxide emissions, 1975–84
Source: US EPA, 1986

sources such as non-ferrous smelters located outside urban areas, as seen in the intermontane region of western United States and in Arizona.

The 1975–84 trend for carbon monoxide at 157 sites shows a 34 per cent decrease for the second highest 8-hour average concentration (figure 8.10). The estimated number of exceedances of the 8-hour air-quality

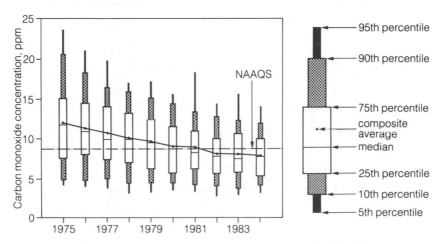

Figure 8.10 National boxplot trend in second highest non-overlapping 8-hour average carbon monoxide concentrations, 1975–84
Source: US EPA, 1986

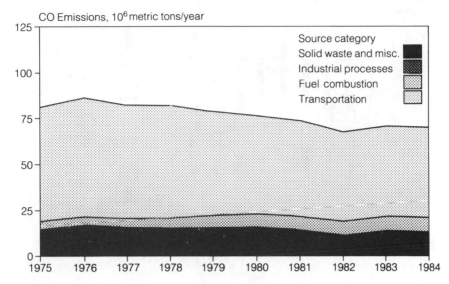

Figure 8.11   National trend in emissions of carbon monoxide, 1975-84
Source: US EPA, 1986

standard decreased by 88 per cent in this same period. National carbon monoxide emissions decreased by 14 per cent between 1975 and 1984 (figure 8.11). When comparing ambient carbon monoxide concentration and emission trends, it is important to recognize that the trends in estimated emissions from motor vehicles involve two components: emissions per vehicle miles of travel, and the number of vehicle miles of travel. Since the early 1970s, carbon monoxide emission per vehicle miles of travel has reduced, but the net effect on national carbon monoxide emissions was dampened by an increase of 30 per cent in vehicle miles of travel between 1975 and 1984. Because carbon monoxide measurements are typically located to monitor potential problem areas, they are likely to be placed in traffic-saturated areas which do not experience increases in vehicle miles of travel. Consequently, the rate of carbon monoxide air-quality improvement reflects the reduction in carbon monoxide emission per vehicle.

Between 1975 and 1984, annual average nitrogen dioxide levels at 119 sites across the nation decreased by 10 per cent (figure 8.12). However, this trend conceals an increase from 1975 to about 1979, a decrease from 1979 to 1983, and then a slight increase again up to 1984. This trend reflects the estimated emissions of oxides of nitrogen from motor vehicles, the source mostly likely to affect the majority of nitrogen dioxide monitoring sites (figure 8.13).

Nationally, the composite average of the second highest daily maximum 1-hour ozone levels, recorded at 163 sites, decreased by 17 per cent

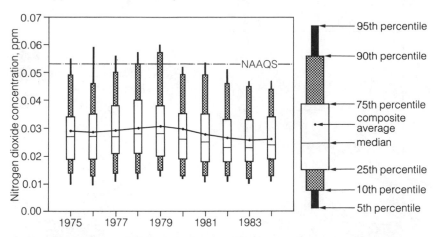

Figure 8.12 National boxplot trend in annual average nitrogen dioxide concentrations, 1975–84
Source: US EPA, 1986

between 1975–84 (figure 8.14). Ozone precursor emissions such as volatile organic compounds (VOC) decreased 6 per cent over this period but this conceals a slight increase from 1975 to 1978, a marked decrease from 1978 to 1982, and an increase from 1982 to 1984 (figure 8.15). In comparing the ambient trends and emission trends, it is important to note

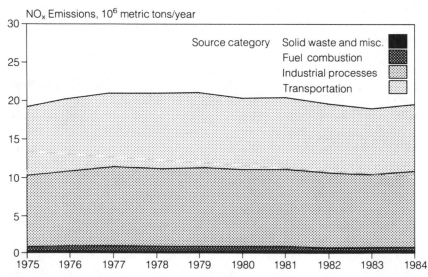

Figure 8.13 National trend in emissions of oxides of nitrogen dioxide concentrations, 1975–84
Source: US EPA, 1986

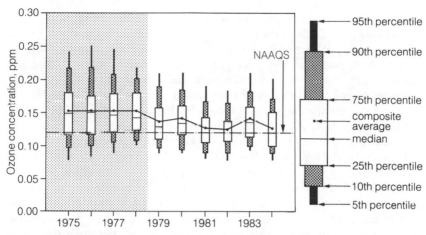

Figure 8.14 National boxplot trend in second highest daily maximum 1-hour ozone concentration, 1975-84. The stippled portion indicates data affected by measurements taken prior to the calibration change
Source: US EPA, 1986

that the apparent improvement in ambient levels from 1978-9 may be partly attributable to a change in calibration procedures. If the post-calibration 1980-4 period is considered alone, then national ozone levels decreased by 9 per cent. The anomalous peak in 1983 was probably caused by meterorological conditions which were more conducive to ozone formation in that year.

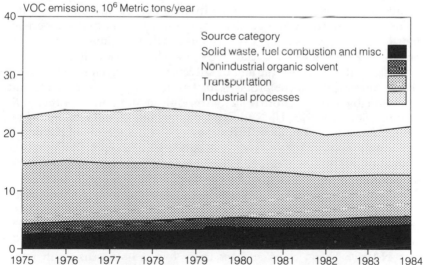

Figure 8.15 National trend in emissions of volatile organic compounds, 1975-84
Source: US EPA, 1986

Although the national improvement in ozone levels is significant, it is less than expected, given the enormous regulation efforts made to tackle this pollution problem. This situation has led to numerous attempts to explain the apparent limited success of pollution-control policies, especially in areas such as Los Angeles where regulatory efforts have been applied more rigorously and for a longer time than in any other area of the United States. Assessment of air-quality trends in that area include Chock et al. (1982), who found no improvement in ozone levels for the 1971–9 period, even after adjustments for changing meteorological conditions from year to year, through Davidson (1986) who found a 15 per cent improvement over the 1976–85 period when a meteorological correction was applied, to the US EPA (1986) who found a decrease of 11 per cent for the 1980–4 period without the application of a meteorological correction. Many researchers point out that what improvements have taken place, if they are indeed agreed to have occurred, were limited, despite substantial reductions in emissions of hydrocarbons and oxides of nitrogen per unit source in the area. Two reasons are adduced to account for this situation: firstly, the precursor emission reductions achieved in the Los Angeles area may have been offset by growth in emissions in the counties neighbouring Los Angeles; and secondly, it may be the case that the oxides of nitrogen have an inhibiting effect on oxidant levels and that the reduction of oxides of nitrogen has counteracted the effect of the hydrocarbon reduction on ozone levels. Certainly, studies have shown that a 50 per cent reduction in emissions of oxides of nitrogen without a concomitant reduction in reactive hydrocarbon concentrations will, in fact, increase ozone formation by about 10 per cent. Moreover, nitric oxide, which constitutes the majority of oxides of nitrogen emissions, is a scavenger of ozone, especially at night. In addition, Walker (1985) points out that maximum ozone levels tend to be associated with moderate rather than high levels of hydrocarbons.

Whatever the reason for the limited success of stringent emission controls in Los Angeles, the ozone problem in that area appears intractable and no reasonable level of effort is likely to result in attainment of the 1987 deadline (Smith, 1981b, US National Commission on Air Quality, 1981); indeed, attainment is unlikely this century. The magnitude of the ozone problem in the Los Angeles area is illustrated by figure 8.16. Although photochemical pollution first became a problem in this area in the United States, it is salient to note that by the 1980s, the problem is as much a problem of the east as of the west of the country (Figure 8.17). Anywhere from between ten and thirty two other cities will miss the 1987 deadline and this is in spite of a relaxation of the ozone NAAQS in February 1979 (Marshall, 1978, 1979). The 1977 Clean Air Amendments required that each NAAQS be reviewed at least every five years to ensure that the standards are based upon the latest scientific evidence (Padgett and Richmond, 1983). Under pressure from the petroleum industry in particular,

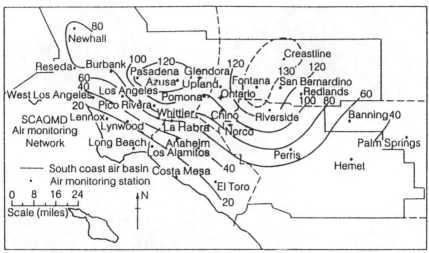

+   Not measured at this location.
-   Less than 12 full months of data.
— — — Intervals of 10 days.
— — — Intervals of 20 days.

Figure 8.16   Number of days on which the ozone standard (daily maximum 1-hour value greater than 0.12 ppm) was exceeded in the South Coast Air Basin in 1984
Source: US SCAQMD, 1985

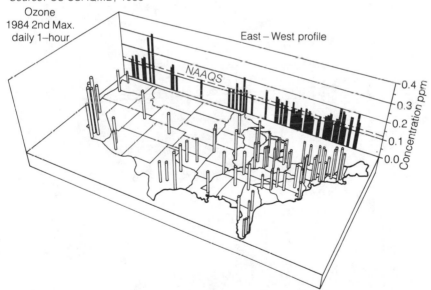

Figure 8.17   Map of the second highest daily maximum 1-hour average ozone concentration at 80 metropolitan areas in 1984. Concentrations are projected onto a backdrop facilitating comparison with the air-quality standard of 0.12 ppm and providing an east-west profile
Source: US EPA, 1986

Table 8.5 Ambient pollution concentrations in 16 metropolitan areas with a population exceeding two million, 1984

Standard Metropolitan Statistical Area	Total suspended particulate (TSP) annual geom. mean $\mu g/m^2$	Sulphur dioxide (SO_2) annual arith. mean ppm	Carbon monoxide (CO) 8-h second maximum value ppm	Nitrogen dioxide (NO_2) annual arith. mean ppm	Ozone (O_3) 1-h second maximum daily value ppm	Lead (Pb) maximum quarterly average $\mu g/m^3$
New York City	64	0.024	15	0.041	0.17	0.91
Los Angeles - Long Beach	86[a]	0.011	19	0.057	0.29	1.02
Chicago	85	0.017	11	0.044	0.15	0.68
Philadelphia	73	0.019	10	0.040	0.20	5.13[b]
Detroit	106	0.014	11	0.025	0.12	0.69
San Francisco - Oakland	59	0.006	8	0.028	0.15	0.50
Washington, DC	64	0.014	14	0.032	0.14	0.49
Dallas - Fort Worth	74	0.005	7	0.016	0.16	1.52
Houston	94	0.010	7	0.029	0.21	0.39
Boston	58	0.016	10	0.044	0.15	0.51
Nassau - Suffolk	49	0.013	10	0.035	0.10	0.67
St Louis	119	0.021	7	0.035	0.17	2.41[b]
Pittsburgh	83	0.035	10	0.031	0.11	0.47
Baltimore	88	0.015	14	0.035	0.15	0.60
Minneapolis - St Paul	75	0.012	13	0.019	0.12	0.65
Atlanta	72	0.009	8	0.027	0.15	0.47
NAAQS	75	0.030	9	0.050	0.12	1.50

[a] Indicates data for 1983.
[b] Reflects the impact of industrial sources.
Source: US EPA, 1986

the EPA was urged to relax the photochemical-oxidant standard (measured as ozone) from 0.08 to 0.16 ppm. However, environmentalists claimed that any relaxation would not provide a sufficient safety margin for people who were most sensitive to smog – children, the elderly, and others with respiratory ailments. The EPA compromised with a revision of the standard to 0.12 ppm, which immediately shifted ten to twenty large cities from the 'non-attainment' to the 'attainment' list, although cities notorious for photochemical smog such as New York, Houston, Philadelphia, St Louis, San Diego, and of course, Los Angeles, remained on the 'dirty' list (table 8.5). Even a decade and a half after the 1970 Clean Air Act, nearly 80 million people are living in areas, primarily cities, that still do not meet the NAAQS for ozone.

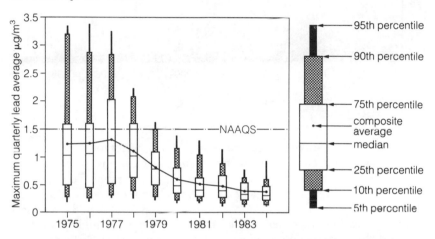

Figure 8.18 National boxplot trend in maximum quarterly average lead concentrations, 1975–84
Source: US EPA, 1986

Major reductions have been achieved in lead levels since 1975, mainly as a result of, first, the progressive reduction of the lead content of petrol, and second, the introduction in 1975 of unleaded petrol for use in vehicles equipped with catalytic control devices (lead 'poisons' the catalyst) which reduce exhaust emissions of carbon monoxide, hydrocarbons and oxides of nitrogen. By 1984, unleaded petrol sales represented about 60 per cent of total petrol sales. For the 1975–84 period, the maximum quarterly average lead levels at thirty six urban sites across the country decreased by 70 per cent (figure 8.18). Over the same period total national lead emissions decreased by 72 per cent (figure 8.19), with the majority of this decrease being accounted for by the transportation sector which fell from a peak of 122,600 tonnes in 1975[1] to 34 700 tonnes in 1984 (table 8.4).

[1] A metric ton or tonne equals 1000 kilograms or 2204.6 pounds, as compared with an imperial ton which equals 2240 pounds.

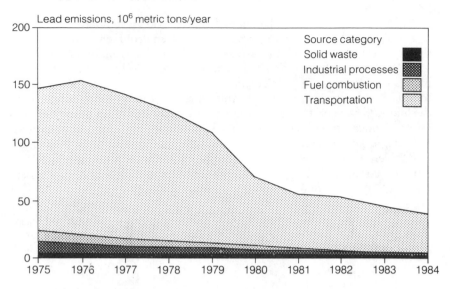

Figure 8.19   National trend in lead emissions, 1975-84
Source: US EPA, 1986

With so much data being reported for so many pollutants and with the news media and control agencies employing diverse terms to describe air quality, some degree of confusion to the public is inevitable. To overcome this communication problem, a health-related Pollutants Standards Index (PSI) was devised for nationwide adoption (Crossland, 1978). The PSI is determined by whichever of the criteria pollutants is highest on that day and the media employ associated terms for various PSI levels such as good (PSI of 0–50), moderate (50–100), unhealthy (100–200), very unhealthy (200–300), and hazardous (exceeding 300). By 1981, 102 urban areas, representing nearly 100 million people, were reporting a daily PSI through the media (US National Commission on Air Quality, 1981).

Considerable progress has been made in improving air quality in the United States during the past few decades but greater improvement is still needed. The rewards are worthwhile both for the health of the population and for the environment. Thus, for example, in 1976 Muskie claimed that '15,000 deaths per year, 15 million days of restricted activity per year, and 7 million days spent in bed' occurred among the population because of air pollution (Lundqvist, 1980). Although the 1970 Clean Air Act originally intended that the country would experience a healthy atmosphere by 1975 (or 1977), it now seems that some areas may not achieve that aim until the turn of the century. Given that many counties will fail to meet the December 1987 attainment deadline for one (or more) criteria pollutants, the EPA has to decide what stance it will take. It can impose bans on the construction of any new major pollution source within such an area and it can impose restrictions on federal funds to the states involved, but it has to do this in a way which will encourage speedier attainment rather than using such measures as a form of penalty or punishment. States will need to be given the opportunity to modify and devise additional control measures within their SIP. It is likely that each state will have to be examined individually for its problems and a new 'deadline' for attainment devised for each case.

8.13 Reauthorization of the Clean Air Act and the Future

During the years following the 1977 Amendments there has been considerable debate, litigation, hearings, and proposals as to what changes were needed when the Clean Air Act came up for reauthorization in September 1981. Industrialists urged sweeping changes to be made to the Act in accordance with their views that the massive cost of meeting clean-air goals has contributed to unemployment, inflation, reduced productivity, and the decline in the country's competitiveness in world trade (US National Environment Development Association, 1981). Their views were shared by the Reagan Administration, which was elected on a platform of economic recovery and deregulation. In contrast, environmental groups argued that tens of millions of people were still subject to unhealthy air and that the Act needed strengthening. It also needed amending to consider the worsening problems of acid rain and toxic chemicals. In the event, opposing viewpoints could not be resolved and the date for reauthorization of the Clean Air Act passed by. The 1970 Act remained in force after that date by way of a continuing resolution.

Generally, the Reagan Administration had planned far-reaching changes to the Clean Air Act. In reality, although some changes have been made, many others have been successfully contested by environmental groups. In 1981, sulphur dioxide emission requirements for major sources were relaxed, based on new dispersion models indicating that larger quantities of pollution could be released without exceeding the ambient air-quality standards. In total, annual sulphur dioxide emission increases of more than one million tonnes were sanctioned (Wetstone and Rosencranz, 1983). In 1981 the Administration required that cost-benefit analysis had to be applied to all major federal regulatory actions in future in order to make measures cost-effective. The EPA has been pressed to continue to experiment with economic approaches to pollution control which allow industry some flexibility.

Not all the changes in regulatory measures during the 1980s have eased or weakened pollution control. Lead in petrol continues to be phased out and the trend towards a ban on leaded petrol (except in special cases) has been accelerated. Although the acid rain problem has yet to be tackled effectively (refer to chapter 11), one of the original causes of acid rain has received attention. Industrial plants with tall stacks are no longer given credit for offsetting sulphur dioxide emissions through the greater dispersion achieved by the tall stack. This is likely to reverse the trend towards the building of taller stacks which began in the late 1960s. Since 1970, 175 industrial stacks over 150 m have been constructed, with seventy four stacks taller than 200 m. Frustration with the slow pace of federal action concerning acid rain has resulted in several states implementing their own acid rain control policies, but federal legislation is the only real solution to the acid rain problem.

Air-quality standards have to be reviewed every five years and this requirement may be seen to benefit industrialists urging relaxation of standards. Thus, for example, the one-hour ozone standard was relaxed in 1979 under fierce pressure from the petroleum industry. However, in 1986 the EPA acknowledged that the health effects from ozone exposure may be more serious than believed when the standard was relaxed and a strengthening of the standard is possible. Furthermore, an eight-hour ozone standard is being proposed (Rombout et al., 1986). Under pressure from the car industry, a revision of the carbon monoxide standard has been proposed which would lower the one-hour standard from 35 to 25 ppm and increase the number of allowable exceedances of the 9 ppm eight-hour standard from one to five (Samberg and Elston, 1982). This latter standard is statistically equivalent to allowing one exceedance of a 12 ppm standard and would shift many metropolitan areas from the 'non-attainment' to the 'attainment' list (table 8.5). This revision in the standard is in line with the car industry's claim that the carbon monoxide emission standard of 3.4 gm/mile is unnecessarily stringent and costly, a view supported by recommendations made by both the National Commission on Air Quality in 1981 and by the National Academy of Sciences in 1980. Other proposed revisions to air quality standards include a stricter TSP health-related standard and the introduction of a standard for fine particulates, since it is these fine particles that are inhaled rather than the non-toxic coarse particulates from roads, construction and agricultural activities which currently distort measurements of suspended particulates. The introduction of a short-term (one-hour) sulphur dioxide standard is also being discussed, and this may help to tackle the problem of acid rain by leading to a reduction in emissions of sulphur dioxide.

The limited success of attempts by the Reagan Administration to ease pollution-control regulations which industrialists deem to hinder economic growth reflects the fact that the public interest in improving air quality is still strong, despite the economic recession which began in the 1970s. Public pressure concerning environmental issues may not exert the same political muscle that it did during the peak of the Environmental Movement in the late 1960s, but it remains strong. Since that time environmental groups have learned to become more effective in challenging any weakening of pollution-control policies or any reluctance by the EPA to protect public health and the environment. Pollution accidents, such as at the Three Mile Island and Chernobyl nuclear reactors and the toxic chemical Bhopal accident, have periodically reinforced public pressure for environmental improvements. The Bush Administration and those Administrations which follow in the 1990s have the challenge of continuing to clean the air and improve the environment while encouraging growth of the economy. Industrialists and environmentalists will continue to argue the point of balance between these two goals. The US EPA already notes that as the national economy has begun to grow again, levels of several pollutants (TSP, sulphur dioxide and oxides of nitrogen) have shown slight increases.

9 Pollution Control in the United Kingdom and the European Community

9.1 Pollution Begins Early in Britain

Air pollution was first recognized as a serious problem during the reign of Edward I (1272–1307) when a Royal proclamation was issued which prohibited the use of sea-coal (coal washed ashore from exposed coal deposits) in open furnaces. The penalty for violating this order was a fine for the first offence and demolition of the furnace for the second. The third offence (assuming one had another furnace) was punishable by death (Martin, 1975). In subsequent centuries less attention was paid to air pollution, despite the efforts of many individuals including the London pamphleteer, John Evelyn, who in 1661 published the very first written report on air pollution entitled 'Fumifugium or the Aer and Smoake of London Dissipated'. Evelyn's schemes to improve the air over London were ignored and it was not until the nineteenth century that the government began its attempts to curb air pollution. The first steps were taken in 1819 when Parliament appointed a committee to investigate how the operation of steam engines and furnaces could be made less prejudicial to public health (UK Open University, 1975).

9.2 The Alkali Acts and the Industrial Air Pollution Inspectorate

The genesis of the modern approach to air-pollution control in the United Kingdom began with the Alkali Act of 1863. At that time the worst effects of pollution were around industrial plants manufacturing alkali, sulphate of soda or potash. The Alkali Act required that 95 per cent of the offensive emissions should be arrested and the remainder be adequately diluted before being allowed to pass into the atmosphere. Following the introduction of this Act, acid emissions from alkali works were dramatically reduced from an annual rate of almost 14,000 tonnes to about 45 tonnes! A second Alkali Act followed in 1874 and this required industrialists to apply the 'best practicable means' to tackle pollution problems (Ashby and Anderson, 1982; UK Open University, 1975).

The Alkali Acts attempt to control industrial pollutants which are specified in a list of 'noxious and offensive gases' associated with scheduled processes ('works'). Over 3000 processes have been scheduled involving over 2000 plants (Sandbach, 1982). The provisions of the Act are administered by a national Alkali Inspectorate. The powers of the Inspectorate have subsequently been increased with the passing of the Alkali, etc. Works Registration Act of 1906 and the Health and Safety at Work etc. Act of 1974. In 1982 the then Alkali and Clean Air Inspectorate was renamed the Industrial Air Pollution Inspectorate but, in April 1987, it became incorporated into a new integrated Inspectorate of Pollution. The Inspectorate must be satisfied that the scheduled process is operated using the best practicable means for preventing the escape of noxious or offensive gases and for rendering such emissions, where discharged, harmless and inoffensive. In practice, this requires the Inspectorate to verify the installation and effective working of approved pollution-control equipment and to indicate minimum heights for industrial chimneys. Emission limits for effluent gases may be set, but except for some standards concerned with total acidity, these are regarded as flexible values to be altered to take account of improving technology and the demands of the public for a better environment (Haigh, 1984; Meetham et al., 1981). All processes that are not in the scheduled lists of the Alkali Acts are the responsibility of local authorities under the Clean Air Acts of 1956 and 1968.

The Inspectorate has frequently been criticized in the past because very few prosecutions of industrialists failing to meet the requirements laid down by the Inspectorate have been made. Between 1920 and 1967, for example, there was a total of only three prosecutions (Wall, 1976a). Even between 1970 and 1975, a period of greater public interest in environmental matters, there were only 20 prosecutions. This reflects the fact that the Inspectorate sees itself as being in partnership with industry, preferring to educate and persuade industrialists to adopt the appropriate pollution-control measures rather than to force them using the threat of prosecution. To some extent this attitude exists because industrialists have not always been as knowledgeable about pollution problems or possible ways of control compared with the Inspectorate, which has a wealth of experience and knowledge gained since it was first formed in 1863 (Gunningham, 1974). The Inspectorate believes that in order to make this partnership work, discussions between the Inspectorate and the industrialist are best kept secret from the public. This has undermined public confidence in the Inspectorate, which appears to have a greater understanding of the

Figure 9.1 The Avonmouth chemical complex has long caused pollution worries for the local community. The establishment of a local industry/community liaison committee by the Industrial Pollution Inspectorate has enabled the community to express its concern

Source: P. Keene

industrialists' concerns than of those of the public and to be insufficiently accountable to the public (Blowers, 1984). However, the situation has improved recently – for instance, the Inspectorate has been actively encouraging the establishment of local liaison committees to discuss problems relating to registered works, and now provides local authorities with details of the 'best practicable means' requirements for individual works. The UK Royal Commission on Environmental Pollution (1984) recommends that this trend towards greater openness be accelerated (figure 9.1).

9.3 London Smogs

The development of air-pollution control in Britain was strongly influenced by the occurrence of a London smog or 'pea-souper' on 5–8 December 1952 (figure 9.2; Ashby, 1975). The 1952 sulphurous smog was characterized by dangerously high pollution concentrations including peak daily concentrations of nearly $4000\,\mu g/m^3$ for sulphur dioxide and $6000\,\mu g/m^3$ for smoke. Smoke refers to suspended particulate matter as measured using the smoke-shade or reflectance method (refer to chapter 2). The smog occurred at a time with temperatures below freezing and this led to a dramatic rise in the number of deaths, especially among the elderly, due to bronchitis, influenza, pneumonia, tuberculosis, and other respiratory illnesses (UK Department of Environment, 1979b). After it had cleared, it was revealed that as many as 4700 people had died as a result of the smog. Deaths due to smog were not to be unexpected: after all, large numbers had died in the earlier London smogs of 1873, 1880, 1881, 1882, 1891 and 1901. What distinguished the 1952 smog from earlier smogs was not simply the greater number of deaths but how the media became involved with the event. Radio and the popular press swiftly reported and commented adversely on what had happened, and public attention therefore became focused on the issue of air pollution (figure 9.3). Consequently, this smog provided the trigger for the public and media to press the government into some form of action concerning the control of air pollution and especially smogs. Public pressure was particularly effective through the campaign adopted by the National Society for Clean Air (formerly the Coal Smoke Abatement Society established in the nineteenth century). As a result, in 1953 the government set up a committee to investigate the nation's air-pollution problems.

Research in London and other British cities had already established that low-level emissions of smoke from domestic open-fires were the major cause of smogs. Some local authorities – notably Manchester in 1946 and Coventry in 1951 – had already attempted to reduce domestic smoke emissions by establishing smoke control areas within which only smokeless

Figure 9.2 The London smog of December 1952 provided the trigger for the public and the media to press the government for stronger air-pollution control legislation
Source: Popperfoto

Figure 9.3 The media image of the London smog of 1952
Source: cartoon entitled 'Smog' by Illingworth (1953), copyright of the *Daily Mail*, London

fuels were allowed to be burned. At the time of the 1952 smog a large number of other local authorities were submitting private bills to Parliament in order to create similar smoke control areas. However, further progress towards the establishment of smoke control areas was halted pending the outcome of the government committee set up in 1953 to examine the problems of air pollution. This committee was established, somewhat reluctantly, to satisfy the public and media pressure for government action. At that time, the government believed that the issue of air pollution had been given undue attention and that as the memory of the December 1952 smog receded, the issue might be forgotten altogether. However, the committee chairman, Sir Hugh Beaver, believed otherwise. He also realized that if the public were without information stimulation during the duration of the investigation, the issue might die and so he released an Interim Report at the end of 1953, close to the anniversary of the 'killer smog'. This report found both the central government and local authorities guilty of negligence in failing to take all possible steps to protect the people of London from a smog disaster, at the same time laying partial responsibility at the doorstep of the people who continued to burn coal (Enloe, 1975). Following the committee's final report in 1954, the government was reluctant to act on the rec-ommendations of the report but was pressed into action by the proposed submission of a private member's Clean Air (Anti-Smog) Bill in 1955 (Gunningham, 1974). This bill was withdrawn on the promise of a government bill being introduced, and so the Clean Air Act of 1956 was passed.

9.4   The Clean Air Acts of 1956 and 1968

The Clean Air act of 1956 granted local authorities the power to designate 'smoke control areas' in which only authorized smokeless fuels (electricity; anthracite; fuels which are inherently low in volatile content, such as gas and oil; or fuels which have had a high proportion of the volatile material removed to enable smokeless combustion, such as types of coke) could be burned. Because these fuels could not be burned in existing domestic heating appliances, the Act provided government grants to be paid to householders to help defray the cost of purchasing and installing new heating appliances (Scarrow, 1972). The Act made smoke-control programmes a matter of local option rather than statutory duty. Given that the local authorities were required to contribute 30 per cent to the cost of the necessary household heating conversions (central government contributed 40 per cent and the property owner 30 per cent), this resulted in the larger, wealthier authorities such as London, Yorkshire and parts of northwest England adopting a progressive programme while other authorities were slow to respond. Cost was not always the most important influence in the pursuit of establishing smoke control areas. In coal-mining areas there was a strong lobby against establishing smoke control areas. Miners received a concessionary coal allowance and designation of smoke control areas would restrict its use. Moreover, Wall (1973) showed that South Yorkshire miners believed that pollution restrictions would reduce the demand for coal, so precipitating redundancies in the coal-mining industry.

Given the regional variation in progress towards designating smoke control areas, central government gave itself the power, through the Clean Air Act of 1968, to require local-government action. Over 300 polluted or 'black' areas were deemed to be priority cases for the introduction of smoke control areas (Burton et al., 1974; Scarrow, 1972). To improve the efficiency with which central government could deal with environmental matters, it created the Department of Environment in 1970. However, the government has not always encouraged the continued establishment of smoke control areas. In 1970–1 many smoke control areas were temporarily suspended because of shortages of smokeless fuels, and this led to a worsening of air-pollution episodes in some urban areas (Elsom, 1979). In the mid-1970s, economic cut-backs by central government led to conversion grants being withdrawn, and between December 1976 and June 1977 a moratorium was imposed upon new smoke control area orders (UK Department of Environment, 1979b). Economic recession during the 1970s caused many local authorities repeatedly to defer the target dates for completion of their smoke control area programme. Nevertheless, by the end of the 1970s, eight million domestic and industrial premises were covered by smoke control orders. During the 1980s there has been renewed designation of smoke control areas as a result of pollution-control policies adopted by the European Community, and also because coal has been identified as an important growth prospect for the UK economy (UK HMSO, 1981). By 1984, almost two-thirds of urban properties were covered by smoke control orders (UK Department of Environment, 1986). In some control areas, illegal burning of bituminous coal persists and this could be increasing winter mean smoke concentrations by 10–20 per cent (Moncrieff, 1985).

The Clean Air Acts of 1956 and 1968 were not only concerned with domestic smoke emissions but also with industrial emissions. Industries not covered by the Alkali Acts, which were therefore the responsibility of local authorities, were pressed to employ the best practicable means of reducing emissions. Unacceptable ground-level concentrations of effluent gases, particularly sulphur dioxide, were controlled by requiring minimum chimney heights. The heights of industrial chimneys were to be determined on the basis of the emission rate of sulphur dioxide (related to the sulphur content of the fuel) when a furnace was operating at maximum, the local topography, and the height of surrounding buildings. Restrictions on smoke emissions from industrial chimneys, especially in the case of black or dark smoke, were enforced using the Ringelmann smoke-shade chart. As with the approach taken by the national Air Pollution Inspectorate, the approach to industrialists by local authorities was one of persuading them to adopt a code of 'good emission conduct'. Even so, prosecutions of industrialists who have contravened the Clean Air Acts have taken place, numbering about 200–300 per annum during the 1970s.

9.5　Air-quality Improvement and Legislation Effectiveness

The government, the media and the public have frequently praised the Clean Air Acts of 1956 and 1968. The First Report of the UK Royal Commission on Environmental Pollution (1971) concluded that 'since the first Clean Air Act became law in 1956 there has been a steady reduction in the emission of smoke and sulphur dioxide into the air over Britain', and it warned that 'the downward trends in smoke and sulphur dioxide pollution are encouraging, but will continue only if there is no relaxation in applying the provisions of the Clean Air Acts . . .' Auliciems and Burton (1973) expressed reservations concerning such strong claims and pointed out that the 1956 Clean Air Act in particular was merely 'swimming along with the social, economic and technological tide'. Indeed, smoke had steadily decreased since the 1920s and probably since the late nineteenth century (Brimblecombe, 1977, 1978, 1982). The traditional British image of a 'cosy coal fire' was changing as affluence followed the postwar depression. A demand for higher heating standards meant that the cleaner and more efficient systems which used solid smokeless fuels, oil or gas (North Sea sulphur-free natural gas became available after 1967) were welcomed. Advertising media stimulated preference for the clean efficient central heating systems. Slum clearance programmes replaced dense terraced housing characterized by multiple low-level emission sources with multi-story dwellings have a central efficient heating system (figure 9.4). Bernstein (1975) and Scarrow (1972) attempted an approximate assessment of the contribution of the Clean Air Acts to the improvement of London air quality. Both concluded that the Acts helped to finance smoke control, through grants for heating appliance conversions, for between only 15 and 30 per cent of London householders. For other less prosperous areas of Britain, especially the north, percentages were higher, but all reveal that social, economic and technological factors were significant in reducing pollution levels.

　Whatever the relative importance of the causes, smoke pollution in urban areas has decreased dramatically since the 1950s (figure 9.5). By the 1970s, urban populations were enjoying the benefits of less suspended particulate

Figure 9.4 City redevelopment, as depicted here in Manchester, helped to reduce urban pollution by replacing dense terraced housing with high rise dwellings which had efficient central heating systems

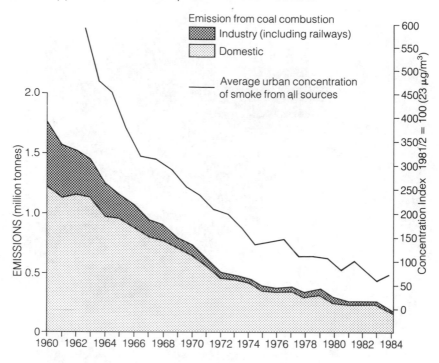

Figure 9.5 Smoke emissions from coal combustion and average urban smoke concentrations in the United Kingdom, 1960-84
Source: UK Department of Environment, 1986

matter in the air in terms of a healthier, sunnier and less foggy atmosphere. In the 1940s and 1950s the number of hours of winter sunshine in city centres such as London and Manchester was only 50 per cent of that of the surrounding rural area, but by the mid-1970s this difference had almost disappeared (UK Department of Environment, 1979b; Tout, 1979). This improvement has taken place partly because of the increase in atmospheric transparency, but mostly through the reduction in the frequency and duration of fogs, as particulate matter encourages the formation of fogs, even at well below 100 per cent relative humidity (Brazell, 1970; Harris and Smith, 1982; Unsworth et al., 1979).

Whereas smoke concentrations in urban areas have decreased dramatically since the 1950s, the decrease in sulphur dioxide concentrations has been less marked. This is due in part to sulphur dioxide being less easy for the public to detect and therefore to complain about. Wall (1974a, 1974b, 1976a) found that in Sheffield between 1949 and 1971, public complaints mentioning only particulates varied between 75 and 96 per cent, compared with only 2–13 per cent mentioning gaseous pollutants at all. The public placed emphasis on curbing smoke pollution rather than invisible gaseous pollutants and so it was not surprising that the Clean Air Acts primarily tackled smoke pollution (especially as expressed in smogs). Nevertheless, even though total emissions of sulphur dioxide have fallen by little more than 30 per cent between 1960 and 1984, average urban sulphur dioxide concentrations decreased by more than 70 per cent. This reduction in sulphur dioxide concentrations is explained by a 75 per cent decrease in low-level emissions (especially from domestic sources) of sulphur dioxide during this period as coal and oil was replaced by sulphur-free fuels such as natural gas (UK Department of Environment, 1986). This highlights that the emissions by power stations, refineries and industry, which are responsible for the greater proportion of sulphur dioxide emissions, add little sulphur dioxide to ground-level concentrations in urban areas (figure 9.6). In addition to emission reduction, the decrease of smoke pollution has helped to reduce sulphur dioxide concentrations by reducing the frequency of fogs which encourage the build-up of other pollutants.

Hydrocarbons and carbon monoxide are two pollutants which are marked by increasing emissions in the United Kingdom (figure 9.7). Motor vehicles and gas leakage are the most important sources. Gas leakage, which refers to the losses during transmissions along the distribution system, accounted for almost two-thirds of hydrocarbon emissions in 1983. Motor vehicles are by far the most important contributor to carbon monoxide emissions, accounting for 84 per cent of total emissions in 1983. In contrast to hydrocarbons and carbon monoxide, emissions of oxides of nitrogen fell by 12 per cent in the period 1979–83. This was mostly due to a decrease in emissions from power stations, whereas motor vehicles continue to increase their relative emission contribution (UK Department of Environment, 1985).

9.6 Control of Pollution Act of 1974

By the early 1970s, ground-level concentrations of suspended particulates had decreased markedly but in some cities, such as London and Sheffield, sulphur dioxide levels remained at a relatively high level. During pollution episodes in London in December 1972 and 1975, peak sulphur dioxide levels significantly exceeded smoke concentrations (Apling et al., 1977; UK Department of Environment, 1979b). Fuel oil was believed to be the major source of the high sulphur dioxide concentrations. Limiting the sulphur content of fuel oil appeared to be the appropriate way to tackle this pollution problem. The then-existing Greater London Council had already set a precedent for this type of pollution-control policy, for special legislation had been introduced in 1972 which required that all new

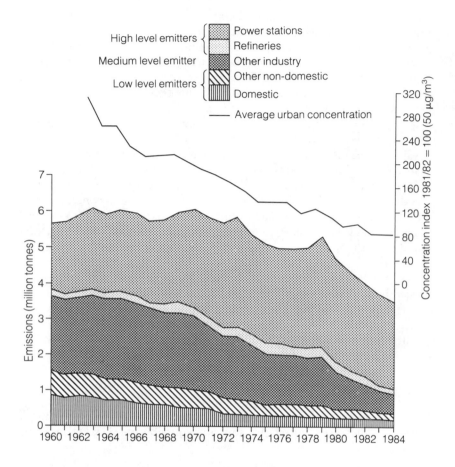

Figure 9.6 Sulphur dioxide emissions[a] from fuel combustion and average urban sulphur dioxide concentrations in the United Kingdom, 1960-84

[a]Emissions from fuel combustion only; excludes emissions from chemical and other processes which probably amount to a few per cent of total emissions from fuel combustion

Source: UK Department of Environment, 1986

oil-fired installations in the City of London were to use fuel oil with a maximum of 1 per cent sulphur content. All existing installations had to comply with this limit by 1987 (Ball and Armorgie, 1983; Masters, 1974). Under section 76 of the Control of Pollution Act of 1974 (sections 75–84 dealing with air-pollution control were not brought into force until 1976), all local authorities are permitted to do as the former Greater London Council had done. To some extent, this provision was introduced in anticipation of a need for it in order to fulfil a European Community (EC) Directive notified in 1975 (Haigh, 1984).

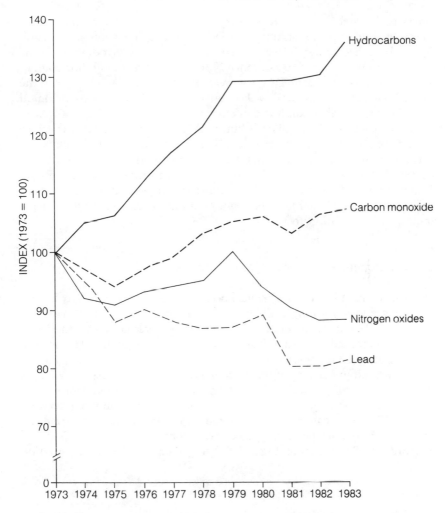

Figure 9.7   Estimated emissions[a] of hydrocarbons, carbon monoxide, oxides of nitrogen and lead[b] in the United Kingdom, 1973-83
[a]The chart is based on constant emission factors and does not take into account possible reductions in emission factors as a result of ECE regulations
[b]From petrol-engined road vehicles only
Source: UK Department of Environment, 1984

   The EC Directive, initiated by West Germany, referred to limiting the sulphur content of light fuel oil or gas oil which describes certain medium distillates used mostly for domestic heating and cooking, as well as 'Derv' for diesel engined motor vehicles. Gas oil differs from fuel oil, which refers to heavier oil used in industry, for heating of large commercial premises, and in power stations. By 1980, the sulphur content of gas oil had to be either below 0.3 per cent or 0.5 per cent, depending upon the situation in which it was used. Since North Sea oil has a sulphur content of about 0.2 per cent, the Directive was not difficult to implement in the United Kingdom, compared with countries using Middle East crude oil with its much higher sulphur content.

Provisions in the Control of Pollution Act also enable local authorities to require industrialists to provide information concerning the emission rates of pollutants from their plants. This information may be disclosed to the public at the discretion of local authorities. The UK Royal Commission on Environmental Pollution (1984) urges that local authorities make greater use of their powers to obtain information and in turn provide increased information to the public.

9.7   Photochemical Pollution and Acid Rain

Since the early 1970s, photochemical pollution has been recognized as both a serious and worsening problem in the United Kingdom. During the exceptionally hot summers of 1975 and 1976, peak hourly ozone levels in London reached 0.150 and 0.212 ppm respectively (table 9.1; Thornes, 1977). London has experienced the worst photochemical pollution in the United Kingdom because of its concentration of the precursor emissions of oxides of nitrogen and hydrocarbons, but on occasions, ozone levels have been simultaneously high in many parts of the country (Derwent et al., 1976). This confirms that polluted air masses containing ozone and other pollutants are being imported from continental Europe (Ball and Bernard, 1978a; Barnes and Lee, 1978; Cox et al., 1975). This situation

Table 9.1   Annual maximum hourly mean ozone concentrations at County Hall, Westminster, in Greater London, 1975-85

Year	Peak value (ppm)	Date of peak	No. of days $\geqslant 0.080$ ppm
1975	0.150	26 June	16
1976	0.212	27 June	26
1977	0.087	3 July	1
1978	0.103	29 July	2
1979	0.153	27 July	5
1980	0.116	27 August	2
1981	0.112	5 September	6
1982	0.091	3 August	1
1983	0.099	16 July	11
1984	0.088	6 July	3
1985	0.098	25 July	3

Source: Ball and Laxen, 1986

indicates that the United Kingdom, although geographically isolated from continental Europe, cannot solve its pollution problems by tackling only indigenous emissions. Co-operation with other European governments is essential if a transfrontier pollution problem, such as photochemical pollution, is to be tackled effectively. Entry into the European Community in 1973 provided the opportunity to begin such co-operation.

Long-range transfrontier pollution of oxides of sulphur and nitrogen leading to acid rain (acid deposition) is clearly a matter requiring European co-operation if it is not to worsen. The effects of acid rain on aquatic ecosystems and forests are only beginning to be highlighted in the United Kingdom. Attempts to tackle the problem of acid rain at the European scale are discussed in chapter 11. However, not every incidence of acid rain in the United Kingdom need be attributed to long-range transport of pollutants from sources entirely in continental Europe. Davies et al. (1984) investigated the occurrence of a distinctive black, acidic snowfall in Scotland on 20 February 1984 in which the snow had a pH level of 3.0. The particulate deposit was very large and consisted of approximately 29 per cent carbon, with the remaining fraction including particles which could be identified as coal fly-ash. The resultant layer of grey snow produced a distinctive smell over the mountainside. Investigations revealed that the heavily polluted air originated from the south-south east – largely, from the industrial sources near Edinburgh and the power stations in the Trent Valley and southern Yorkshire. The pollution was probably transported at an altitude of 1000–1500 m in association with a stable atmospheric layer whereupon particles were efficiently scavenged by snow over the Cairngorm Mountains (figure 9.8).

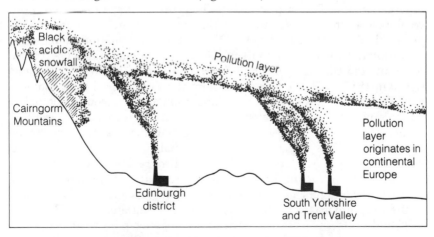

Figure 9.8 Suggested pollution sources contributing to the black, acidic snowfall in Scotland on 20 February 1984
Source: modified from Davies et al., 1984

One pollution problem related to the transport of the precursors of acid rain concerns summer visibility degradation. Lee (1983a, 1985) has shown that summer visibility throughout rural southern England declined by between 20 and 30 per cent in the period 1962–73 due to increased sulphates, and to a lesser extent nitrates, in the atmosphere. Increased emissions of precursor pollutants from motor vehicles were initially thought to be a significant factor in this problem but the effects of the 1973 oil crisis suggested otherwise. This crisis reduced oil consumption in industries and power stations throughout Europe, so reducing the production of sulphates. Coinciding with this change in emissions, visibility improved markedly after 1974. This improvement in visibility took place even though vehicle fuel consumption continued to show an upward trend through 1974, thus suggesting that pollutants from motor vehicles were not the main influence on visibility.

9.8   Entry into the European Community

In 1973 the United Kingdom, along with Denmark and Ireland, entered the European Community (EC) which at that time consisted of Belgium, West Germany, France, Luxembourg, Italy and the Netherlands. Greece acceded in 1981, followed by Spain and Portugal in 1986. Although it is common to refer to the European Community in the singular, there are in law three separate Communities: the European Economic Community (EEC), the European Atomic Energy Community (EURATOM) and the European Coal and Steel Community (ECSC), each established by a different treaty. The institutions of these three Communities are now combined so that to refer to the Community, in the singular, reflects practical reality (Haigh, 1984). Of the three Communities, it is the European Economic Community which is the most important. The Treaty of Rome (the constitution for the EEC) was drafted in the 1950s, and signed in 1957, well before the environment was conceived of as a subject requiring systematic international collaboration, and consequently it makes no reference to the environment as a community matter. It is purely an economic treaty, devoted to encouraging 'harmonious development of economic activities and a continuous and balanced exapansion'. In the development of an environmental control programme, the Community decided not to amend the Treaty of Rome, but rather, to find justification for such a programme in the existing provisions. As a result, the purpose of the EC's Environment Action Programmes (initiated in 1973, 1977, 1982 and 1986), are presented in the context of attempting to reduce unfair competition and to promote harmonious trade. This is pertinent since pollution-control measures can account for between 20 and 30 per cent of the costs of some industrial production processes, and if pollution-control requirements varied between countries, this would constitute a barrier to fair competition (Johnson, 1979).

Although the Commission of the European Community would prefer to introduce pollution-control regulations directly into national laws of each Member State, in practice it issues Directives in which the terms are required to be incorporated into national legislation before they are effective (Farquhar, 1983). The formulation of a Directive is usually marked by bargaining and compromise between Member States and a lengthy succession of modified draft Directives. Many Directives which began as drafts carrying specific provisions have been reduced to the status of framework Directives containing no more than a general and largely qualitative list of requirements, accompanied by a reference to the fact that further subordinate but more specific Directives would be issued at a later stage. In practice, this has proved little more than a delaying tactic to cover up the fact that while everyone agrees with general statements that it is right to prevent and reduce pollution, there is no agreement on the details regarding how it is to be done or, indeed, why it should be done. Nationalism often surfaces – that is, a wish by Member States not to have to change existing and, in some cases, long-established policies. Given that Member States have the right to veto any proposal, this can lead to progress towards pollution control falling victim to the convoy system – that is, the slowest dictating the speed of the whole group.

9.9   European Community Air-quality Standards

During the 1970s, the EC became convinced that there was sufficient medical evidence to justify proposing health protection air-quality standards ('limit values') for various pollutants. In July 1980, the EC issued a Directive (80/779/EEC) setting out the maximum concentrations of smoke and sulphur dioxide permitted in urban areas (table 9.2). In recognition of the synergistic effect of these two pollutants on health, the health protection standard for sulphur dioxide is dependent on the level of smoke present: the more smoke that is present on a particular time-scale, the less the amount of sulphur dioxide allowable. Other pollutants were to be tackled later.

This strategy represented a radical change in the approach to pollution control adopted in the United Kingdom. Consequently, strong opposition was expressed to the proposed introduction of air-quality standards (UK Royal Commission on Environmental Pollution, 1976). This was one reason for the proposed EC Directive taking four years before being finally adopted. British strategy had relied heavily on the willingness of industrialists to employ the 'best practicable means' to reduce pollution rather than to apply coercive penalties (Burton et al., 1974). In contrast

Table 9.2   Air-quality standards (limit values) for sulphur dioxide and smoke as specified in the European Community Directive (80/779/EEC)[a]

Reference period	Limit value for SO_2	Associated value for smoke	Absolute limit value for smoke
Year	80	>40 (34)	
(Median of daily			80 (68)
means	120	≤40 (34)	
Winter	130	>60 (51)	
(Median of daily			130 (111)
means October – March)	180	≤ 60(51)	
Day	250[b]	>150(128)	
(98th percentile			250[b](213)
of daily means)	350[b]	≤150(128)	

[a]Smoke concentrations (suspended particulates) are expressed in $\mu g/m_3$ according to the OECD Smoke Calibration Curve, with the equivalent values for the BSI Smoke Calibration Curve given in brackets. Conversion from OECD to BSI smoke measurements can be made by multiplying OECD values by 0.85.
[b]Not to be exceeded for more than three consecutive days.
Source: Compiled from information supplied by the Commission of the European Communities, Brussels, Belgium

with this approach, many other Member States of the EC had serious doubts as to whether a system seemingly based on voluntary compliance could ever be adequate. They believed more in having clearly defined and mandatory air-quality standards (Levitt, 1980). The concept of air-quality standards was not entirely new to the United Kingdom but the 'standards' which existed were only non-mandatory guidelines or reference levels such as those adopted by the Greater London Council in the mid-1970s. It was in this form that the United Kingdom was willing to adopt air-quality 'standards'.

Member States were given until 1983 to comply with the air-quality standards, with the tactics for implementing the standards being left to the individual states. However, at the outset it was realized that some areas could not be brought into compliance by this date so they were given until April 1993 to achieve compliance. The EC Directive emphasized that the implementation of the standards in urban areas must not lead to a deterioration of air quality in the 'clean' regions; as far as possible, compliance with the standards must be achieved by reducing emissions and not by wider dispersal of pollutants in the environment (Johnson, 1979). It is not clear precisely what legal rights the EC has to enforce such a Directive and what penalties are to be given for non-compliance. The United Kingdom has not introduced new legislation to ensure compliance with the air-quality standards: it believes that the Clean Air Acts of 1956 and 1968, the Alkali etc. Works Regulation Act of 1906, the Control of Pollution Act, 1974 and the Road Traffic Act of 1972 are adequate.

The Clean Air Acts give local authorities power to control domestic and industrial smoke and empower the Secretary of State for the Environment to direct local authorities to submit and implement smoke control programmes. Local authorities are currently being urged to complete the smoke control programmes they began in the 1950s or 1960s but which were given low priority in the 1970s, as smoke pollution had already fallen sharply and because the 1970s was a time of economic restraint. Given that the reason why all of the areas in the United Kingdom could not make the 1983 compliance date was because of smoke alone, and that the smoke was primarily from domestic coal-burning, then smoke control area designation seemed an obvious choice of control measure. The Alkali Acts empower the Industrial Air Pollution Inspectorate to control emissions from registered chemical and industrial processes. The Control of Pollution Act empowers the Secretary of State for the Environment to control the composition of motor fuel and the sulphur content of fuel oil, and also empowers local authorities to obtain information about air pollution. The Road Traffic Act 1972 empowers the Secretary of State for Transport to regulate the construction of vehicles to avoid smoke and other emissions, and gives powers for authorized examiners to check vehicles (Haigh, 1984).

The EC requires Member States to set up an adequate network of air-pollution monitoring stations to supply information to ensure compliance with the air-quality standards. Since 1961, the United Kingdom has had an extensive smoke and sulphur dioxide monitoring network, the National Survey of Air Pollution, which at times has consisted of as many as 1200 sites. From April 1982 this network was drastically rationalized to form the United Kingdom Smoke and Sulphur Dioxide Monitoring Network (Basic Urban Network) of 150–175 sites (Elsom, 1982; Handscombe and Elsom, 1982). The reason for this rationalization was that smoke and sulphur dioxide concentrations had fallen so much that costly monitoring was no longer necessary. An alternative reason is that such a dense network increases the number of individual sites which may indicate failure of one or more of the EC standards. As the UK Greater London Council (1979) stressed, 'it would be unfair if we and other European cities with extensive monitoring networks had to undertake greater expenditure to reduce pollution levels than less carefully monitored cities'. In spite of this comment British cities are currently required to maintain some 295 sites, in addition to the rationalized national network of 150 or so sites, in urban areas where there is a possibility that the EC standards may be approached or exceeded in the next few years.

The adoption of the EC health protection air-quality standards for smoke and sulphur dioxide in urban atmospheres represents a major step towards harmonizing pollution control within the EC. It opened the door for further air-quality standards to be subsequently defined. Standards for lead and nitrogen dioxide have since been added and EC Directives are

Table 9.3 Air-quality standards (guide values) for sulphur dioxide and smoke as specified in the European Community Directive (80/779/EEC)[a]

Reference period	Sulphur dioxide ($\mu g/m^3$)	Smoke ($\mu g/m^3$)
Year (Arithmetic mean of daily means)	40 to 60	40 to 60 (34 to 51)
Day (Daily mean value)	100 to 150	100 to 150 (85 to 128)

[a]Smoke concentrations (suspended particulates) are expressed in $\mu g/m^3$ according to the OECD Smoke Calibration Curve and with the equivalent values for the BSI Smoke Calibration Curve given in brackets.
Source: Compiled from information supplied by the Commission of the European Communities, Brussels, Belgium

being prepared for carbon monoxide and hydrocarbons. Once all of the legislative, administrative and monitoring machinery is established for air-quality standards, it provides the possibility of making the current standards more strict in future years. Already the EC has specified non-mandatory and more stringent secondary air-quality standards ('guide values') to protect the environment (table 9.3). Rosencranz (1982) viewed the smoke and sulphur dioxide limit values as lenient and largely symbolic: by the time the Directive was applied, less than 5 per cent of the Community's land mass exceeded the standard. However, strengthening of these standards is likely in the future. Already, concern for acid rain is leading to the consideration of upgrading the sulphur dioxide guide values to become the enforceable limit values. If this action was indeed taken, it would pose enormous problems for some countries, given that there are several urban-industrial areas which already have an extension to 1993 in order to achieve the current limit values.

The trend towards setting stricter air-quality standards was evident in the setting of the nitrogen dioxide standard in 1985. The air-quality limit value, expressed as a 98 percentile of recorded hourly mean values over the year, is $200\,\mu g/m^3$ (approximately equivalent to a median value of $75\,\mu g/m^3$). This value appears particularly strict given that the lowest level at which nitrogen dioxide has been demonstrated to have an effect is $940\,\mu g/m^3$. The air quality standard thus includes a four- to five-fold safety factor. Some argue that this is unnecessarily strict and that a standard of, say, $300\,\mu g/m^3$ would have been adequate, which is nearer the top of the range suggested by the World Health Organization (UK House of Lords Select Committee on the European Communities, 1984). An air quality guide value of $135\,\mu g/m^3$ for the 98 percentile and $50\,\mu g/m^3$ for the 50 percentile of hourly mean values is also specified (Laxen, 1985).

9.10   Motor Vehicle Emissions

Control of pollution from motor vehicles is achieved through regulation of the composition and content of motor fuel and by the use of vehicle construction and vehicle-in-use standards. Regulation of motor fuels has been concerned mostly with lead in petrol and sulphur in diesel fuel. The legal powers to regulate motor fuels were included in the Control of Pollution Act of 1974, which allowed the government to implement the EC Directive concerning lead in petrol in 1978. In regulating the emissions of particulates and gases from motor vehicles, the EC is strongly influenced by the views of the United Nations Economic Commission for Europe (ECE). This organization has a Working Group developing guidelines on emissions from petrol-engined vehicles which can be observed by all countries in Europe. To date, the EC has adopted the ECE guidelines in the form of Directives. The Directives also ensure that member States do not set more stringent limits than those laid down in the ECE guidelines, and barriers to trade are thereby prevented (Haigh, 1984). EC Directives on carbon monoxide and unburnt hydrocarbons were introduced in 1970 and limits for oxides of nitrogen were added in 1977. Since then, the limits for all three pollutants have become more strict with successive Directive amendments. The latest proposal is to reduce emissions of these pollutants to the very strict levels of Japan and the United States by 1995.

The 1975 London air-pollution episode highlighted how particulate emissions from motor vehicles are now of more importance in that city than emissions from coal burning in domestic premises. At the time, with 92 per cent of Greater London being covered by smoke control orders, many were surprised that smoke concentrations could still reach undesirable levels. It was subsequently shown that particulates from vehicles contributed on average about 75 per cent of 'dark smoke', but during pollution episodes this reached as high as 90 per cent (Ball and Hume, 1977; McGinty, 1977). The significance of this contribution may increase if the trend towards greater use of diesel engines for propulsion of light- and heavy-duty vehicles continues. Diesel-engined vehicles emit about four times the mass of particulate matter per unit weight of fuel consumed as compared with petrol-engined vehicles using leaded fuel. The significance of this difference is magnified, given that petrol engines using unleaded fuel typically discharge little particulate matter (Ball, 1984). If smoke concentrations are to be further reduced in London, it is the vehicular source of particulates which has to be controlled. Pressure for tighter control of emissions from diesel engines is likely to intensify, given the increasing concern shown by the British public for the perceived health risks of vehicle emissions and the soiling effects of road vehicles (Ball and Caswell, 1983).

9.11   Lead-in-petrol Debate

Lead in petrol first became an EC issue as long ago as 1971 as a result of proposed German legislation severely restricting its use (Haigh, 1984). After many years of discussion, the matter eventually became the subject of a 1978 Community Directive which set the maximum permitted lead content of petrol sold within the Community at 0.40 grams per litre (g/l), to be met by 1981. Member States may set an upper limit between 0.40 g/l and 0.15 g/l, but they cannot insist on less than 0.15 g/l. Ireland was given a ten-year extension for compliance. The limit of 0.15 g/l was chosen because it is near the lowest level usable in existing petrol engines without special adaptations. The inclusion of the lower limit ensured that no barriers to trade in motor vehicles would be created by any one Member State insisting on unleaded petrol.

In the United Kingdom the amount of lead permitted in petrol has fallen from 0.84 g/l in 1972 to 0.40 g/l in 1981 and 0.15 g/l in December 1985. This has led to emissions of lead from petrol-engined road vehicles decreasing markedly since emissions peaked in 1973 at 8.4 million tonnes (figure 9.7). Lead in petrol became a national issue in the United Kingdom in 1982 when an environmental pressure-group called CLEAR, the Campaign for Lead-Free Air, was established (Wilson, 1983). This pressure group, chaired by Des Wilson, was highly effective in mobilizing public and media involvement and in lobbying British and EC politicians. The public campaign and the recommendations of the ninth report from the Royal Commission on Environmental Pollution eventually resulted in 1983 in the British government's requesting that the EC Directive be amended so as to remove the minimum limit contained in it. The British government accepted the Royal Commission's target date of 1990 for the introduction of unleaded petrol throughout the EC but urged an earlier date to be adopted if that was possible (UK Department of the Environment, 1983). In Europe, unleaded petrol in fact means a lead content of between 0.01 and 0.02 g/l, which is similar to the United States limit of 0.0135 g/l (Lubinska, 1984).

At the same time as the United Kingdom was arguing for the introduction of unleaded petrol for health reasons, West Germany was pushing for the introduction of unleaded petrol as soon as possible for other reasons. Increasing damage to their forests had alarmed the German public and they believed that motor vehicle exhaust emissions were to blame, at least in part, for the forest damage. To reduce exhaust emissions the Germans intended to introduce catalytic converters on new models of cars from 1989. Since these do not work efficiently with leaded petrol, then an early ban on lead in petrol was needed. Denmark, Luxembourg and the Netherlands joined West Germany in their willingness to require the introduction of unleaded petrol by 1986, while others argued that this date should be optional. In 1984, the proposed Directive concerned with phasing lead out of petrol was the subject of 158 amendments, and no final draft was in sight. Given such delays in producing a final draft explains why one Member State, West Germany in this case, threatened to break EC regulations and act unilaterally because the issue was seen to be of such importance in its own country. Under this pressure a final Directive was agreed which allowed unleaded petrol to be introduced in 1986 but did not *require* it to be introduced until 1989.

The concentration of lead in air is also the subject of an EC Directive which requires that Member States ensure that the annual mean concentration of lead in air does not exceed $2 \mu g/m^3$ by 1987, or by 1989 in areas given an extension. This Directive thus introduced another pollutant for which an air-quality standard applies and adequate national monitoring networks have had to be developed to ensure compliance. A separate Directive also requires monitoring of blood lead levels in the population.

9.12 The Impact of EC Membership on Pollution Control in the United Kingdom

For a hundred years the United Kingdom followed an emission-control strategy referred to as the 'best practicable means' approach. In the 1970s, the EC rejected the suitability of this strategy for the whole Community and instead adopted the air-quality management approach to pollution control. By 1983 the United Kingdom had to comply with smoke and sulphur dioxide air-quality standards, and several other standards were expected to follow. Other EC Member States were subjected to the same imposition but in no other country did it represent such a radical departure from previous practice as in the United Kingdom.

It was not surprising that the UK resisted the adoption of air-quality standards. Indeed, the concept has been debated by the UK Royal Commission on Environmental Pollution in 1976, with the recommendation that standards should be rejected in favour of non-mandatory air-quality guidelines. Nevertheless, by 1980 the United Kingdom had accepted the introduction of standards; but the resistance it had exhibited, and the delay for which it was responsible in the implementation of the EC Directive, raised questions concerning the United Kingdom's commitment to international air-pollution control. In other forms of pollution, especially water quality, Britain has found herself as the voice of dissent among the Community. In recent air-pollution issues, the United Kingdom's intitial views concerning the EC's proposals for emission restrictions on large combustion plants has again placed it out of line with majority opinion. Given the country's long experience and achievements in air-pollution control, it is disappointing not to see Britain taking a more positive role in the development of EC policy (Haigh, 1984).

Many people believe that it is time for the United Kingdom to take a leading role in air-pollution issues within the Community. There is much to be gained from taking the initiative in policy-making since it then gives that country a greater chance of influencing the direction and detail of Community policy. By taking a more positive and leading role, the UK could not only regain its traditional reputation as a country which initiates pollution-control policies (e.g. the Alkali Acts in the nineteenth century and the Clean Air Act in 1956), but it could also gain commercial and political advantages too. There are commercial advantages in pursuing high environmental standards, as this will help to sustain an internationally competitive pollution-control equipment industry. A country which can maintain a position in the forefront of pollution control will have considerable opportunities for the export of appropriate technology and equipment, provided others follow suit with similar abatement policies (UK Royal Commission on Environmental Pollution, 1984). The EC is learning this lesson the hard way as it moves towards the strict motor vehicle exhaust emissions of Japan and the United States. This move will provide the Japanese with a great commercial advantage in Europe since its vehicles are already fitted with catalytic converters for its home market and that of the United States.

Membership of the EC seems to have been a helpful stimulus to environmental protection in Britain. Air quality in the United Kingdom has, or will have, improved as a consequence of EC Directives. The EC seems to have injected a greater sense of urgency into improving air quality than was apparent in Britain in the 1970s and 1980s. The designation of smoke control areas, initiated following the Clean Air Acts, was being given a low priority by the mid-1970s, but the introduction of air-quality limit values for smoke and sulphur dioxide has given renewed impetus for completion of smoke control area programmes. Research into the nature and effects of a wide range of pollutants has been stimulated in anticipation of many other air-quality standards being proposed in the future. Another practical effect has been to maintain in place an air-pollution monitoring network more extensive than there would otherwise have been, and monitoring a wider range of pollutants (Haigh, 1984).

Since the 1970s, an increasing number of pollution problems involve pollutants which cross national boundaries and whose control depends on international co-operation. For that reason alone, the United Kingdom can but gain in the long run from working within the EC. Moreover, since the EC, with its 12 Member States and 320 million population, represents an economic unit of considerable power, and one which is able to legislate, this means that it can set the agenda for other international gatherings and exert considerable influence (Haigh, 1984). The influence of the EC is already being recognized in international pollution-control discussions: for example, the Vienna Convention for the Protection of the Ozone Layer (refer to chapter 11) was negotiated and signed by the Commission of the EC on behalf of its Member States in 1985. Through the EC, therefore, the United Kingdom has the potential to have a much greater influence on what action is taken concerning international and global pollution problems.

10 Approaches to Pollution Control in Socialist Countries

10.1 Socialism: Balancing Economic Growth and Environmental Protection

The pollution problems which characterize Western societies may be believed by some to arise because economic production is based principally on private ownership and profit-making. Pollution occurs because selfish greed for profits results in the social costs of pollution being ignored. In a socialist country there is no contradiction between the interests of society and the interests of individuals; all of industry is government-owned and economic production is controlled directly by the state, and the government is therefore both the principal polluter and the protector of the environment. This should mean that pollution can be controlled more readily in socialist countries and that industrial and urban growth is more compatible with notions of conservation, environmental protection, and improving the quality of life (Sandbach, 1980). This is the theory, but is it true in reality? Can a socialist state strike a better balance between the need for economic growth and the need to protect the environment than can be struck by a capitalist state? This chapter considers the air pollution problems and pollution-control policies of the Soviet Union and the People's Republic of China in an attempt to answer this question.

10.2 The Soviet Union

10.2.1 Rapid Industrial Growth: a Central Government Priority

Industrialization came late to the Soviet Union and as a consequence, as most developing countries, a high priority has been given to the pursuit of the rapid industrial development and modernization of its economy, especially during the past 50 years. It appears that this desire for economic growth has overridden environmental considerations: only since the early 1970s, for example, has significant attention been paid to the pollution of the air, unless it was obviously objectionable (Enloe, 1975). Pollution tended to be referred to mainly in terms of something that happened in capitalist countries because of uncontrolled exploitation of workers and resources.

Recently, there has been an increasing awareness of the high and ever-rising cost to the nation, in terms of health and resources, of air pollution. This resulted in the singling out of pollution control for its own special high budget (RBL 1000 m or £700 m) in the 10th Five Year (1976–80) Plan (UK Department of Trade and Industry, 1983). This contrasts with earlier five-year plans which showed little concern for the environment despite existing declarations in the Soviet constitution. The effectiveness of this recent commitment to pollution control has yet to be fully assessed, although one review of the 1976–80 Five Year Plan pointed out several ministries which had fulfilled air-pollution abatement plans by between only 48 and 82 per cent (Pryde, 1983).

10.2.2 Air-quality Legislation

Since 1951 the Soviet Union has followed the Air-Quality Management strategy for air-pollution control and 114 air-quality standards have been designated (Izmerov, 1973). These standards, known as 'maximum permissible concentrations' (MPCs), are claimed to be the world's toughest air-quality standards. In fact, the standards are so strict that they are frequently not met. Full enforcement of the standards is an impossible task because these standards are numerous and in most cases extraordinarily stringent. Thus, for example, the 24-hour MPC is $50 \, \mu g/m^3$ for sulphur dioxide, $50 \, \mu g/m^3$ for TSP and $85 \, \mu g/m^3$ for nitrogen dioxide (Glass, 1975). The standards are based on research on health effects alone, without regard to considerations of available control technology, economic feasibility, or the ability adequately to measure these concentrations in practice. When a standard is currently unattainable, it represents a direction for future enforcement or a guideline for future research in control technology (Glass, 1975). The standards are based on toxicological evidence alone and are considerably more strict than most international standards, which are based on epidemiological studies. In practice, Komarov (1980a) explains that any air-quality reading exceeding an MPC by between one and five times represents a potential danger to human health; readings over ten times greater constitute an 'immediate threat' to human health; and readings over 25 times greater constitute conditions of 'extreme hazard' to health.

In the past, major stumbling blocks to effective pollution control in the Soviet Union have been, first, the low standing in the ministerial hierarchy of pollution-control agencies, and second, the separation of central (or state) and republic responsibilities towards economic growth and

Figure 10.1   Main areas of industrial pollution in the Soviet Union
Source: modified from UK Department of Trade and Industry, 1983

environmental protection. Large-scale industries such as iron and steel, chemicals, mining and power, which have a great potential for environmental pollution, are widely distributed throughout the republics of the Soviet Union (figure 10.1), but they are accountable only to the central government. The environmental regulations of the republic in which the industry is located (all republics wrote environmental regulations between 1957 and 1964 which instructed factories to take steps to meet the MPCs) are not binding on the industrialists. In effect, this means that pressure on industrialists to fulfil production quotas set by a central directive overrides consideration of pollution control. Even when a dispute concerning a pollution issue gets discussed at the ministerial level, it is usually settled by political weight – and party connections – of the contending organizations, rather than on the basis of the objective merits of the case (Enloe, 1975; Kelley et al., 1976).

Taken from the perspective of the factory director, there have been few incentives that have been concerned with the environmental effects of production. Like his superiors at the ministerial level, the director has been judged in terms of gross output or the profitability of his enterprise (Kelley et al., 1976). Until profit-oriented reforms were enacted in the 1970s, the sole criterion for success was the total output of the factory; nothing else mattered – not the quality of the product or sales, and certainly not the environmental consequences of production. Under the profit system, the enterprise is evaluated by a profit standard, but this has not notably heightened the environmental consciousness of factory managers, for pollution-control expenditures are regarded in most cases as unproductive ones that diminish the firm's profit. It is not surprising that pollution is a serious problem (Komarov, 1980a, 1980b).

Only recently has an attempt been made to resolve this conflict of responsibilities between state and republic. In 1980, the Air Quality Law was passed (this complements an earlier similar law on water quality) and it became the first national law on air pollution. The Air Quality Law takes a wide view and quotes plans for 'protecting the air basin as part of the state plans for the economic and social development of the USSR, by the exercise of state control of air protection, and also by the establishment of ceilings on permitted discharges from the possible sources of pollutants'. The basic concept for the control of air pollution is 'to fix quotas' (emission standards) for the discharge of pollutants at every enterprise, in respect of every source of pollutants. This represents a radical departure from the traditional air-quality management approach to controlling air pollution. While the Law stresses the requirement for clean air over the whole country, when it comes to specific restrictions, these appear to apply more to newly-built factories, power stations, etc. than to existing ones. As one might expect, emission standards for older factories are less strict than those for newer ones because of the costs involved in reaching the emission standards. There appear to be considerable difficulties of installing modern environmental control equipment in old factories built before the revolution, especially in the industrial areas of the Southern Urals. The effectiveness of the Air Quality Law remains to be seen. Much will depend on the fixing of emission 'quotas' and the rigour with which they are enforced. Previous experience would indicate that while tough standards may be fixed, in practice many exceptions will be allowed. Production quotas have always had implicit priority and this is unlikely to change (UK Department of Trade and Industry, 1983).

Overall responsibility for all pollution control, monitoring, research and development is associated with the State Committee for Hydrometeorology and Environmental Monitoring (Goshydromet), set up in 1978. This is a potentially powerful body whose chairman is a member of the USSR Council of Minsters. The Environmental Protection Department of the State Committee for Science and Technology (SCST) appears to advise Goshydromet, as does an Interdepartmental Scientific and Technological Council jointly set up by the SCST and the USSR Academy of Sciences. As all industrial and most agricultural enterprises in the Soviet Union are state controlled, the position of Goshydromet in relation to the large number of ministries responsible for individual industries is not completely clear. It appears to have the power to close down polluting factories and to require local authorities to conform to laid-down standards, but it is not known if or when it has used this power. This power cannot be applied directly; it involves the Council of Ministers of the relevant individual republic of the Soviet Union in addition to the local authority. A rapid decision is therefore unlikely. Local authorities do not appear to have much power to enforce centrally laid-down regulations, but they are encouraged to bring polluters to the notice of Goshydromet.

10.2.3 Air Quality: Worse or Better than Western Countries?

Although recent air-quality data is not readily available, some subjective assessment of current air quality can be made, especially by considering the information on air-pollution levels which we do have from the 1960s and from the work of Komarov (1980a, 1980b). In the mid-1960s, the withering of trees and the yellowing of pine needles became commonplace in industrial areas, and in the Urals, tree-kills up to a 10 km radius around some factories were the norm (Kelley et al., 1976; Pryde, 1972). It was concern for analogous effects occurring in the residential areas that resulted in the introduction of the concept of 'air-pollution buffer zones'. Highly polluting industries are required to be at least 2000 m from the nearest residential community, but in practice, this planning regulation is frequently violated (Kelley et al., 1976). Some industrial cities have experienced 'deadly smogs' reminiscent of London in the 1950s (Mote, 1974).

Today, pollution still remains a serious problem in some localities. Thus, for example, in the copper-nickel smelting cities of Noril'sk in Siberia's Taymyr Peninsula, and Monchegorsk in the Kola Peninsula, pollution concentrations exceed MPCs by 20–25 times (Bond, 1984; Komarov, 1980a). Sulphur dioxide and other acidic effluents from industrial processes, power stations and refineries are a considerable problem. Most indigenous oil and much Soviet coal is of high sulphur content. Attempts have been made to try to use lower sulphur content fuels in or near cities (reserving the higher sulphur content fuels for more remote areas), but how widespread desulphurization is adopted is difficult to assess.

Recently, environmental deterioration has attenuated because of a slowdown in industrial expansion since the late 1970s and because resource conservation has become much more important following the worldwide energy crisis in 1973 (Pryde, 1983). The Soviet Union claims a reduction of almost 15 per cent in the discharge of air pollution in the past five years, though the original levels of air pollution were not made public (UK Department of Trade and Industry, 1983). At the United Nations Multilateral Conference on the Environment in 1984, a Soviet official claimed that the USSR had reduced sulphur dioxide emissions by 2 million tonnes over the period 1980–4. Although large emission-sources produce severe air pollution, much publicity is given to the assessment that the total amount of air pollution in the Soviet Union is only between one-quarter and one-half that of the United States – or at least this was the case in the early 1970s (Mote, 1974). This does not imply that pollution control is more effective in the Soviet Union: rather, it is because of the much lower population density of the Soviet Union as a whole, together with the lower level of economic development of the country and the relative scarcity of motor vehicles.

Not all cities experience poor air quality. Moscow suffered a deterioration of its air quality in the 1930s and 1940s, but subsequently became the focus of a massive air-pollution clean-up campaign. More than 700 major polluting industries were moved out of the city and thousands of purification installations were introduced, together with district central heating for homes using natural gas (Glass, 1975; Izmerov, 1973). However, even this showcase of environmental action suffers from national problems such as the excessive exhaust emissions and noise from motor vehicles. Badly adjusted and old diesel engines are widespread on trucks and buses. Komarov (1980a) claims that the MPCs of carbon monoxide in some areas of Moscow are sometimes exceeded by between 20 and 24 times, and are constantly exceeded by 10–13 times. This statement needs qualification, as the Soviet Union carbon monoxide standard is remarkably low compared with even the World Health Organization's long-term goal. Whereas the World Health Organization specifies $10 \, \text{mg/m}^3$ for a 8-h period and $40 \, \text{mg/m}^3$ for a 1-h averaging period, the Soviets employ $1 \, \text{mg/m}^3$ for a 24-h standard and $3 \, \text{mg/m}_3$ for a short-term (20-minute) standard (Glass, 1975).

Motor vehicle engine pollution-control measures appear completely inadequate in the Soviet Union. Low compression engines are used almost without exception and most of the petrol sold is of a surprisingly low octane number. Seventy-two or seventy-six octane petrol is widely available, though in recent years 93 or, in some larger cities, 95 octane has become increasingly used as more private cars appear on the roads (UK Department of Trade and Industry, 1983). In the larger cities only unleaded petrol is being made available. Even though the number of cars per head of population is still low in Western terms, it has been growing rapidly as cars become more readily available. The difference in motor vehicle numbers and pollution produced is quite astonishing: for example, in 1978, whereas the Soviet Union had 18–20 million vehicles, the United States had 110 million, yet Komarov (1980a) points out that Soviet vehicles pollute the atmosphere by four times as much as their American counterparts. As the Soviet Union attempts rapidly to expand its vehicle manufacturing industries, Komarov predicts that vehicles will be responsible for almost 70 per cent of the air pollution by 1990.

10.2.4 Pollution Control and Monitoring Technology

Air-pollution-control equipment appears generally inadequate in the Soviet Union. Pollution-control devices are often not finished by the time factories open. Older industrial plants often have simple and ineffective control equipment (Goldman, 1970; Kelley et al., 1976; Pryde, 1983). There is a chronic shortage of Western currency with which to purchase foreign control-equipment. Not surprisingly, a considerable effort is being expended in an attempt to raise the level of the country's control and monitoring technology. The current inadequacy of pollution-control equipment does not seem to delay the establishment of new factories, despite the safeguards which are claimed to be in force. Thus, for example, before a factory commences operation, inspectors are supposed to certify that control equipment is adequate and operational and that the MPCs will not be exceeded. Once operations have begun, the inspector is required to check the equipment periodically in order to ensure efficient operation. In practice, delaying the opening of a factory, or shutting it down because of failure to meet these pollution-control requirements, rarely takes place because of the national importance afforded to economic production. Komarov (1980a) reports that over 100 cities have an average level of 10 MPCs for noxious gases in the atmosphere and that in those cities that are centres for metallurgical and chemical industries, average levels may exceed 100 MPCs.

The inadequacy of pollution-control equipment has implications not only for pollution within the Soviet Union but also for pollution outside the country. Evidence from satellite images of emission plumes from the smelter stacks of industrial complexes, such as the one at Noril'sk, points to the Soviet Union being one of the major contributors to episodes of 'Arctic haze' as far away as central Alaska (Shaw, 1982). Given the limited availability of effective pollution-control equipment to tackle internal pollution problems, it is unlikely that equipment will be installed for the sake of an international problem such as 'Arctic haze'.

Although Goshydromet is currently still in the process of developing a network of monitoring stations throughout the country, the pollution monitoring system in some cities is claimed to be particularly advanced. In Tashkent, for example, the monitoring of pollution concentrations and atmospheric conditions resulted in an 'alert' being proclaimed on 12 June 1984 and in prompt action being undertaken to reduce emissions (*Soviet Weekly*, 20 October 1984). Industrial plants were instructed to reduce emissions to prearranged lower levels, asphalt boilers on construction sites were shut down for several hours, and road-works raising dust were allowed only at night until the pollution episode had abated.

10.2.5   Public and Media Influence on Policy-making

There is, of course, no overt environmental lobby of the type found in most Western countries. The average Soviet citizen is not likely to make him or herself conspicuous by complaining about pollution. Even so, environmental groups do exist in the Soviet Union, such as the All-Russian Society for the Conservation of Nature, which has 19 million members in the Russian Republic and six million in the Ukraine, with lesser numbers in other republics. Most of the members are, however, school children and the organization has little political leverage at the national level (Kelley et al., 1976). Of far greater importance at the national level seem to be the increasing numbers of scientific and medical advisers to the central government who are beginning to comment more frequently in private about the personal hazards and economic losses associated with pollution, especially of the air (UK Department of Trade and Industry, 1983).

Air and water pollution caused a loss to the nation's economy of 5 per cent of the national income in 1980 (Komarov, 1980a). This increased awareness of the effects of pollution is beginning to be reflected by the media, which is very significant, given that the media are state-controlled. In 1986, a new television series openly dealt with questions on environmental protection. However, the lengths to which some people have to go to be able to criticize environmental protection in the Soviet Union is illustrated by the case of Komarov. Boris Komarov is a pseudonym used by a Soviet author (Zeev Wolfson) who later emigrated to Israel, but who wrote, and subsequently smuggled out, a damning assessment of environmental conditions in the Soviet Union in 1978 (Komarov, 1980a, 1980b). His assessment of atmospheric pollution was indeed highly critical, as illustrated by the statement: 'That clean air has become a rarity is a fact even our lungs recognize – it is obvious'.

The degree to which the Soviet people react to air-pollution issues may depend in part upon the strength of their underlying Marxist belief and traditional revolutionary viewpoint that nature's resources are to be exploited for the interests of the people; the problems of pollution will be solved by improved technology. The people have long been promised a better material standard of living, so there is underlying pressure from the people for economic growth and especially for the availability of good-quality consumer goods. For a long time the people have been deprived of, or made to wait on, lengthy lists for consumer goods. This situation may mean that some environmental deterioration from new factories producing consumer goods would be acceptable. Soviet consumer materialism, like its capitalist counterpart, could result in a worsening of air-pollution problems in the near future. Of particular impact would be a substantial increase in the production of motor vehicles scheduled to go into private ownership rather than government motor pools.

The absence of a powerful environmental lobby is very noticeable in relation to the Soviet Union's development of commercial nuclear power (Pryde, 1983). To conserve fossil fuels, as well as to reduce conventional pollutant emissions, the Soviet Union accelerated its development of commercial nuclear power stations during the late 1970s and early 1980s, such that the proportion of the nation's electricity which nuclear power provided rose from almost 6 per cent in 1980 to 10.3 per cent in 1985. The Soviet Union did not admit to reservations concerning the safety of its nuclear power stations, and at the time of the Three Mile Island reactor accident there was considerable coverage in the Soviet media designed to demonstrate that such a failure could not happen in the Soviet Union and that the local population was safe from Soviet nuclear power stations.

Unfortunately, and ironically, in 1986 the world's worst nuclear reactor accident occurred at Chernobyl, releasing 1000 times more radionuclides than the Three Mile Island accident (refer to chapter 5; Hawkes et al., 1986). This was not the first accident involving ionizing radiation: reactor accidents had happened before, such as that in 1969 (Trofimenko, 1983), whilst in December 1957 a major chemical explosion involving nuclear wastes required the population to be evacuated from within 200 km of the point of the explosion at Kyshtym on the eastern flanks of the Urals. This latter disaster only came to light following a publication by an emigré Soviet scientist (Medvedev, 1979). Although Chernobyl became more openly discussed by the Soviet leadership and media than any previous environmental disaster in the Soviet Union, Chernobyl is unlikely to delay significantly the Soviet nuclear power expansion programme (figure 10.2).

10.2.6 Assessment of Current and Future Importance of Pollution Control

The pursuit of economic growth still remains the Soviet Union's priority and pollution control is subservient to that goal. The development and technological modernization of its industry is regarded as the key to the nation's independence and influence in the world. Recently, events such as the passing of the Air Quality Act in 1980 and the increasing comment on the economic losses caused by air pollution suggest that pollution control is increasingly coming into favour. However, strong legislation and strong declarations have been made before, only to suffer from a reluctance to translate these pronouncements into firm commitment and action. What can be achieved is highlighted by the clean-up of Moscow, but the need to elevate this city to a pollution-control showcase suggests the inadequacy and ineffectiveness of routine pollution-control procedures. The recent upgrading in importance of pollution control through the

Figure 10.2 By 1986, the Soviet Union had 51 nuclear power stations in operation, 34 under construction and another 39 planned. Despite the Chernobyl disaster, the expansion of nuclear energy production is unlikely to be slowed. At this Armenian nuclear power station, at Metsamor, a team of researchers check on radiation levels

Source: Novosti Press Agency

establishment of Goshydromet holds promise, but if it is to achieve the results planned for it, there will be a conflict of interests between it and the various Ministries of Industries within the Soviet Union. In the past, pollution-control agencies have seldom won their arguments with the pro-industry groups and there is no clear indication that this situation will change in the near future.

To conclude that pollution control is of limited importance and effectiveness in the Soviet Union is disappointing because the Soviet political system does have some potential strengths for tackling environmental protection. The presence of both a highly centralized administrative system and nationwide economic planning gives Soviet leaders the ability to affect virtually all aspects of governmental and economic activity. A manipulative and highly effective communications system further provides a mechanism both to educate and mobilize mass support through the media; and an extensive network of Communist Party cells or party representatives within virtually all governmental, economic, and social organizations provides an effective transmission belt for explaining and enforcing policy. Its pattern of social and cultural norms emphasizes obedience to formal authority, facilitating the task of enforcement of policies and laws. Its collectivist socialist ethic could easily provide the frame of reference for measuring the social costs of environmental deterioration and assessing the social and economic trade-offs necessary for effective control programmes (Kelley et al., 1976).

All of these factors would mean that, if the Soviet leaders so wished, pollution control could be implemented vigorously and effectively. Unfortunately, given the overwhelming pressures to modernize the economy and devote greater attention to meeting the consumer demands of the society – to say nothing of the political pressures to preserve the existing dominance of a pro-industrial and pro-growth elite – it is unlikely that Soviet leaders will elevate pollution-control policies to a much higher level than they currently occupy.

10.3   The People's Republic of China

10.3.1   Politics and Pollution

Up until the late 1970s, China claimed that its planned socialist economic system produced far fewer environmental problems than countries with capitalist economies. Capitalist countries, it was claimed, exploited, polluted and even destroyed the environments of their own and other countries. In contrast, China adopted policies and practices which gave careful consideration to the need to protect the environment. Chinese propaganda emphasized that the problems of urbanization were less severe because 80 per cent of the Chinese population lived in rural areas. Problems arising from heavy industry were fewer because the population was urged to develop small-scale local industries. Emphasis was placed upon the recycling of wastes, or what the Chinese call 'multi-purpose utilization' in production (Hall, 1977; McDonald, 1975). This reflected the Chinese concern for curbing the 'three wastes' – namely, waste-water, waste gas and solid wastes. Not only was recycling beneficial to industries faced with scarcity of resources but it also reduced the quantity of waste products discharged into the environment. Wastes were also used for energy production. Biogas (mainly methane) could be produced through the fermentation of garbage, human and animal excrement, leaves and straw in sealed digesters.

While such policies and practices are environmentally commendable, and similar to those advocated by some environmental pressure-groups in Western countries, current assessments suggest that they have been of only limited success in China (Smil, 1980a, 1980b, 1984). The proportion of rural population may be high but China still has the greatest number of cities inhabited by over one million people of any country in the world. The rapid pace of industrialization and urbanization has far exceeded the capacity of Chinese ways of, and efforts in, recycling wastes. Small-scale industries waste fuel because of boiler inefficiencies, and the installation of adequate pollution-control equipment in such small-scale industries is too costly even if it were available – which it is not. Currently, seven million biogas digesters have been built in the countryside, providing fuel for 35 million people, but biogas provides only a very small proportion of the total fuel requirement in China.

Today, the People's Republic of China openly admits that it faces pollution problems as serious as any other nation intent on rapid economic development. Not only does it acknowledge the seriousness of its environmental problems, but it blames some of these difficulties on mistakes made during the Cultural Revolution (1966–76). It further admits that pollution problems were virtually ignored until the 1970s. Currently it recognizes that economic growth can be harmed by pollution problems; and it acknowledges that 'preservation and improvement of the environment is an important condition to ensure a more effective and sustained development of the economy' (Qu, quoted in Swannack-Nunn et al., 1979). It recognizes that air pollution may harm important national resources such as food crops, forests, soils, lakes and the workforce, thereby hindering economic development. Pollution may be costly to control, but China now accepts that its effects are too costly to ignore.

Although China is now tackling pollution problems, it believes that the rate of economic and industrial development should not be reduced for the sake of environmental protection (Qu and Li, 1984). It argues that the environment should be protected and improved during the course of 'harmonious' economic development. Economic development is a state priority and it has set the target of quadrupling the value of gross industrial and agricultural production by the year 2000, thereby placing China in the 'front ranks of the countries of the world' in terms of gross national income and the output of major industrial and agricultural products. Of course, per capita income would, as the Chinese acknowledge, remain low compared with that of major industrial countries. In 1982 the population passed 1008 million and the one-child-per-couple policy attempts to ensure that population in the year 2000 does not exceed 1200 million.

10.3.2   Problems of Fuel and Technology

The fundamental reason for many of China's air-pollution problems is its great reliance on coal for industrial processing, power generation, cooking and space heating. In 1982, coal accounted for 74 per cent of the total energy consumption, and it is estimated that coal consumption in cities will increase by 30 per cent by the year 2000. The quality of much of the coal used is relatively low (Shen, 1984): the coal generally has a high ash content and only about 30 per cent of the coal is washed before use, whereas physical cleaning of coal (washing) could reduce ash content by 50 per cent and sulphur content by 30 per cent (Zhao and Sun, 1986b). Raw coal is used as a household fuel either in the form of lump coal or more often as honeycomb briquettes compacted from poor quality coal dust, the combustion of which is poor. Although the sulphur content of Chinese coal is relatively low, the lack of processing produces an average sulphur content of end-use-coal of 2–3 per cent, and in south China it may even reach 10 per cent (Kinzelbach, 1982, 1983). Not only is the quality of the coal poor but the combustion efficiency of its industrial plants is very low (30–40 per cent). Chang (1985) reports that the improvement of the combustion efficiency in China's numerous outdated thermal power plants and 200,000 industrial boilers and furnaces could reduce particulate and gaseous emissions by as much as 50 per cent. This would also lead to considerable savings in coal consumption.

Households contribute to pollution because of the dominance of coal, especially in the form of coal-dust briquettes used for heating and cooking in small stoves. Combustion in these stoves is inefficient; only 10–18 per cent of the chemical energy in the coal is converted to useful thermal energy, compared with 80–90 per cent for large boilers used in Western industrialized nations (Chang, 1985). Some reduction in the use of coal in households is becoming evident, and increased use of liquefied petroleum gas for cooking is being reported. Power plants are being built to provide more residents with electricity. The development of natural gas resources, such as the development on Hainan Island, will result in a greater availability of a cleaner fuel. Nevertheless, a reduction in pollution from households must rely in part on the progress of economic development and the associated social changes. From the experience of most developed countries, it is apparent that improved standards of living create a demand for cleaner and more efficient heating and cooking systems. In addition, urban renewal programmes are resulting in individual houses with poor heating systems being replaced with high-rise dwellings having heating systems supplied by an efficient collective heating system.

Although China may emphasize its adoption of small-scale industries, heavy industries are of great importance to the nation: indeed, the first Five Year Plan (1953–8) was based on the Soviet model of development, with priority given to heavy industry. Manufacturing industries are mostly located in the coastal cities such as Shanghai and Tianjin and in the northeast region (formerly Manchuria), although most large cities are developing new industrial areas. Many of the heavy industries initially established, such as the vast Anshan iron and steel complex, lacked pollution-control equipment. Any attempts at pollution control in the early years were not helped by the Soviet withdrawal of technicians and aid in 1960. Although China currently sets out to incorporate pollution-control equipment in new industrial plants, it lacks the foreign currency to purchase state-of-the-art foreign technology. Chang (1985) claims that pollution-control equipment is almost non-existent and that the development of a pollution-control industry is just beginning. Kinzelbach (1983) claims that only 1 per cent of China's power-plant capacity has electrostatic precipitators, while it is reported by Chinese officials that among China's 400,000 enterprises, only 200–300 electrostatic precipitators are in operation, and those are of uncertain efficiency (Weil, 1981). Obligatory installation of electrostatic precipitators would virtually eliminate the gross black smoke still polluting the atmosphere in many Chinese cities.

10.3.3   Pollution Levels

Air pollution is an obvious problem in most cities in China, especially in the northern cities in winter. Information on pollution levels in cities is limited, but table 10.1, compiled by the US Environmental Protection Agency's delegation to China in October 1979, highlights the seriousness of air pollution and the possible associated damage to human life and working days (Chang, 1985). The level of suspended particulates reflects the different types of industries in different cities. Wuhan is a steel city in the middle of the Yangtze valley, and has the highest death rate believed to be caused by suspended particulates; while Guangzhou, which is a city of light industries, records the lowest death rate of sampled cities. Even so, pollution in Guangzhou is very evident to the visitor. The panoramic view of the city which one can enjoy at 6.30 a.m. vanishes by 7.00 a.m.

Table 10.1   Suspended particulate matter concentrations in selected cities in China in 1979 and public health statistics

	Average suspended particles in 24 hours ($\mu g/m^3$)	Maximum content of suspended particles in 24 hours ($\mu g/m^3$)	General absenteeism of industrial workers per annum (1000 days)	Number of deaths believed to be caused by air pollution per annum
Beijing	80	160	25,000	850
Shanghai	150	200	40,000	1300
Wuhan	170	400	22,000	3500
Guangzhou	190	190	20,000	700

Source: Chang, 1985

as hundreds of boiler stacks begin to spew forth plumes of black smoke that quickly merge to form a dirty smoke-haze.

In general, levels of suspended particulates are high in most Chinese cities, but they are highest in the north (figure 10.3; table 10.2). Sources of particulates include not only combustion products from industry and households, but also exposed soil in and around urban areas. Large areas of China are covered by fine loess soils which are readily transported by winds. Vast tracts of land suffer from severe soil erosion such that dust

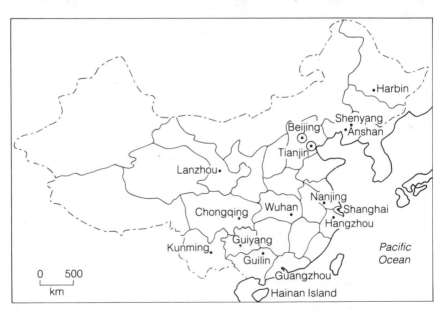

Figure 10.3   The People's Republic of China: locations referred to in the text

Table 10.2   Daily concentrations of total suspended particulate and sulphur dioxide in selected Chinese cities

Location	City	July average	December average	Annual average
		Total Suspended Particulates, μg/m^3		
North	Shenyang	324	769	470
North	Beijing	165	585	403
South	Shanghai	161	322	244
South	Guangzhou	133	194	174
		SO$_2$, μg/m^3		
North	Shenyang	9	691	132
North	Beijing	53	388	158
South	Shanghai	53	121	65
South	Guangzhou	48	53	52
South	Chongqing	280	610	430
South	Guiyang	347	409	413

Source: Zhao and Sun, 1986a

storms are frequent. Daisey et al. (1983) found that dust storms increased the TSP concentrations in Beijing by a factor of two or three during the spring of 1981. On average, wind-blown dust accounts for 60 per cent of TSP concentrations in summer and 40 per cent in winter in northern cities (Zhao and Sun, 1986a). As the air contains high concentrations of alkaline (desert dust) particles, rainfall acidity may reach pH levels of 7.0–7.5. Nevertheless, an increasing incidence of slightly acidic rain-storms (pH 5.0–5.5) is being monitored in Beijing, while in parts of southern China, where the particulate content of the air is lower, rainfall pH is 4.5–5.5 (Jernelov, 1983). Extensive urban redevelopment and construction programmes also add to the particulate burden of urban atmospheres. Given the unpleasantness to pedestrians and cyclists of dust raised by passing vehicles, it is not an uncommon sight to see a few people wearing face masks during the rush hours to reduce inhalation of the dust.

Vehicle pollution has not reached the seriousness of the problem facing most developed nations simply because there are fewer vehicles in Chinese towns and cities. In the early 1980s, China had only 1.5 million civilian motor vehicles, of which just 180,000 were passenger cars and the rest trucks and buses. Only since 1986 has private ownership of cars been allowed. Although vehicle emission standards exist, they are considerably higher than those employed in the West and Smil (1984) reports that Chinese-made cars produce 15–55 times higher hydrocarbon emissions than comparable Japanese cars. As the numbers of vehicles increase in Chinese towns and cities, vehicle pollution will become a serious problem, as will photochemical smog.

Table 10.3   Noise levels in selected Chinese cities and cities in Western countries

City	L_{NP} (dB)	TNI[a] (dB)
Guangzhou	95	108
Hangzhou	111	143
Chongqing	106	132
Wuhan	109	136
Nanjing	105	127
Harbin	95	137
Hong Kong	80	87
London	74	68
Madrid	89	89
Medford, Mass.	69	69
New York	88	96
Rome	90	88

[a]The value of TNI is derived from the equation $TNI = L_{90} + 4d - 30$ db, where L_{90} denotes the minimum noise level recorded more than 90 per cent of the time; L_{10} denotes the minimum noise level recorded 10 per cent of the time; and $d = L_{10} - L_{90}$. L_{NP} is derived from the equation $L_{NP} = L_{50} + d + d_{60}^2$ db.
Source: Chang, 1985

No visitor to Chinese cities can fail to recognize noise as an environmental problem. Table 10.3 compares noise levels between selected Chinese cities and cities in Western countries (Chang, 1985; Ko, 1978). The principal sources of noise are industrial machinery, the extensive construction programme being undertaken, the large number of small tractors and lorries which travel back and forth between the city and the countryside 24 hours a day, and the all-too-frequent use of vehicle horns. Noise control is being attempted by moving industry away from residential areas, by screening factories with trees (Noble, 1980), and by banning the use of horns in parts of cities. However, progress in improving the situation is slow.

With limited resources available for pollution control throughout China, it is not surprising that national priority has been given to reducing pollution in selected cities such as Beijing and Tianjin, and in a few tourist centres such as Guilin. Even this limited commitment will be costly, given that Smil (1984) reports that 70 per cent of the industrial stacks in Guilin had no fly-ash control in 1980 and that the pollution had caused severe damage to the surrounding karst landscape. However, for most cities significant reductions in air pollution will take much longer, and like many other developing countries in Asia striving for industrialization and material improvement, a clean and quiet urban environment may still be considered a 'luxury' in measuring the quality of Chinese life (Chang, 1985).

10.3.4   *Pollution-control Framework*

Environmental management only became a major issue in China following the United Nations Conference on the Human Environment at Stockholm in 1972. China held its first national conference on environmental protection in 1973. In recognition of the mounting complexity of environmental problems, a General Office of Environmental Protection (also known as the State Council Environmental Protection Office) was established under the State Council in 1973. In 1982 a new ministry the Ministry of Urban and Rural Construction and Environment Protection, was created which now co-ordinates and supervises the country's environmental protection (Qu and Li, 1984).

China follows the Air Quality Management approach to pollution control (Yao, 1980). Since 1963 a large number of largely academic air-pollution standards have been in circulation throughout China, but they have varied in both origin and geographical extent of application, and most have been approximate guidelines rather than firm limits (UK Department of Trade and Industry, 1984). However, the need for stronger measures was recently acknowledged, and in April 1982, national ambient air-quality standards, to be applied under the Environmental Protection Law of the People's Republic of China of 1979, were finally published (Table 10.4). Although the standards are in most cases less strict than equivalent standards in the West, they are presented in some detail and represent a major step forward. What is now needed is a clear and detailed statement concerning the strategy to be adopted to ensure that urban areas achieve these air-quality standards and the timetable allowed for compliance with the standards. Much greater effort needs to be placed on both monitoring and enforcing the standards. Achieving these standards is likely to require a substantial amount of time and resources. Only recently 290 monitoring stations have been set up and they suffer from incomplete sets of equipment and inadequate numbers of qualified personnel. Whereas the state sets out the policies for pollution control and the standards of air quality, the task of enforcement is left to the local environmental protection bureaux or offices. It appears that these provincial or municipal environmental-protection bureaux vary considerably in their effectiveness.

Although as of yet, a strategy for achieving and maintaining compliance with the air-quality standards may be unclear, some action is very evident and is also widely publicized. A system of fining persistent industrial polluters in some urban areas is frequently mentioned. Fines are set a little higher than the costs of pollution control so as to provide an economic incentive to industries to introduce pollution control earlier rather than later. The amount of the fine depends upon the quantity and concentration

Table 10.4　Ambient air quality standards in China

Pollutant	Sampling time	Concentrations, mg/m_3		
		Class 1[a]	Class 2[b]	Class 3[c]
All atmospheric particles (TSP)	Average over one day	0.15	0.30	0.50
	Maximum at any time	0.30	1.00	1.50
Dust ($< 10\mu m$)	One day	0.05	0.15	0.25
	Max	0.15	0.50	0.70
Sulphur dioxide	Average over one year	0.02	0.06	0.10
	One day	0.05	0.15	0.25
	Max	0.15	0.50	0.70
Nitrogen oxides	One day	0.05	0.10	0.15
	Max	0.10	0.15	0.30
Carbon monoxide	One day	4.00	4.00	6.00
	Max	10.00	10.00	20.00
Ozone	Average over one hour	0.12	0.16	0.20

[a]Class 1 standards apply in state designated nature conservation beauty spots, historic sites and places of convalescence.
[b]Class 2 standards apply in residential areas, mixed commercial, traffic and residential zones, cultural areas, historic sites, villages and other zones, as designated in municipal plans.
[c]Class 3 standards apply in cities, towns and industrial zones with relatively high atmospheric pollution, city traffic intersections, major arteries and other zones.
The standards refer to separately published detailed guidance on sampling techniques.
Source: Compiled from information supplied by the Environmental Protection Office of the People's Republic of China, Beijing

of pollutant discharged. The director of the offending factory must personally pay between 2 and 5 per cent of the sum. The workforce is also penalized since the accumulation of large fines for a particular industrial unit diminishes the amount of the year-end bonus in which each worker shares. Eighty per cent of the fine is spent directly on purchasing pollution-control equipment for the offending factory, while the remaining 20 per cent goes to the government. Despite the publicity given to fines imposed on some industrial plants, observations in Chinese cities highlight the fact that many factories emitting black smoke are currently tolerated rather than fined. This reflects the national importance given to economic and industrial growth, as well as a widespread shortage of effective emission-control equipment.

The use of fines is just one economic approach to pollution control; another relates to the incentives which are given to industries which use waste gas, waste liquid and other residues as their main material. They qualify for tax reductions or exemption, or for special considerations in price policies. They may even keep profits instead of turning them over to the state (Qu and Li, 1984).

Progress towards air-quality improvement relies on co-operation between the Environmental Protection Office, industry and the public. Propaganda is used extensively throughout China to educate the public on the desirability of improving air quality (figure 10.4; Guo, 1981; Qu and Li, 1984). To publicize the implementation of the Law of Environmental Protection, a 'propaganda month' was launched in March–April 1980, with propaganda relating to environmental protection being issued through the press, films, television and photographic exhibitions. More than 690 articles on environmental protection were carried in national and provincial newspapers (Qu and Li, 1984). Propaganda posters continue to be displayed on notice-boards in urban areas. Experience in many developed countries highlights the significant progress that can be made towards air-quality improvement when there is widespread support from the public and from pressure-groups. Neighbourhood committees have established supervisory groups for environmental protection in some

Teacher: What's the chief element in the air?
Students: The bad smell from the factories.

Figure 10.4   Cartoon propaganda used to educate the public on the desirability of improving air quality
Source: Beijing Review, 15 April 1985, p. 29 (cartoonist: Cheng Jinguo)

cities and have participated in legal cases against factories which pollute residential areas.

10.3.5 Industry and Pollution

One major industrial-pollution problem currently being tackled is that of the enormous number of small inefficient boilers. In Beijing alone, for example, there are 4260 small boilers with a capacity of less than 1 tonne of steam per hour and these contributed about one-third of the sulphur dioxide ground-level concentrations (Siddiqi and Zhang, 1984; Zhao and Sun, 1986b). Heat-energy utilization is typically less than 30 per cent in such units and emission of black smoke is a common sight (Smil, 1984). Actions being taken to solve this problem include the replacement of the inefficient boilers with large modern boilers, the increased fitting of industrial boilers and kilns with dust removers as this equipment becomes available, and the conversion of boilers to a gas-firing basis. The current development of large-scale natural gas resources will both accelerate the latter trend and encourage the development of cleaner industries in general.

Industry has been encouraged to consider methods other than the fitting of expensive emission control equipment to stacks in order to reduce pollution near factories. One method investigated by the South China Botanic Research Institute is for factories to make use of what are termed 'anti-pollution plants'. Plant species are brought from far afield and their tolerance to pollutants is examined in laboratory fumigation experiments. Suitable species are then planted around factories to reduce the dispersion of particulates and gases away from the area using the trees and shrubs as a physical barrier, to absorb the gaseous pollutants, and to collect the gaseous and particulate pollutants on the increased surface area offered by the plant foliage, especially when the foliage is wet. This approach to pollution control also introduces pollutant-tolerant plant species into an area where other species had previously died, and the trees help both to screen obtrusive factories and to reduce the noise near factories. Li (1979) reported the success of planting 17,000 trees and shrubs around the Guangzhou Chemical Works in southern China, from which waste chlorine was emitted. The tree planting was claimed to have reduced chlorine concentrations near the factory by 20 per cent in two years. Plant species selected for use in southern China to date include mango, alpine fig, beef wood, and oleander for chlorine pollution; rose apple, almond, and mulberry for sulphur dioxide pollution; and fan palm, *Ficus Elastica* and a number of others for hydrogen fluroide[1] pollution. This approach to

[1]Fluorides are produced during the manufacture of steel, phosphate, bricks, tiles and glass. They are a heterogeneous group of compounds formed from the highly reactive, non-metallic gaseous element known as fluorine. Animals eating fluoride-polluted herbage develop fluorosis

tackling pollution around factories and industrial parks has the potential for much greater national and international adoption.

Industry is publicly encouraged to find ways to reduce pollution by regular newspaper features publicizing the pollution-control successes achieved by exemplary industrial plants. Thus, for example, the Anshan Iron and Steel Company is said to have reduced the volume of dust falling in the vicinity of its factory in north-east China from 280 tonnes/sq. km per month in 1977 to 130 tonnes/sq. km per month in 1981 (UK Department of Trade and Industry, 1984). Reductions in the effects of pollution plumes from industry are also being attempted through appropriate land-use zoning of urban areas (Qu, 1982). Like many other countries, China has experienced the adverse health effects of valley or basin smogs into which heavy industry has added dangerous substances. Lanzhou, in the Gansu Province, for example, experienced a 'killer smog' in 1977. Consequently, industrial districts are being established outside urban areas, but this appears to be a long-term strategy of pollution control as industries emitting black smoke are frequently in evidence within residential areas (figure 10.5).

10.3.6   China's Commitment to Pollution Control

At the end of 1983 it was reported that China spent between about 4 and 5 billion yuan a year on environmental protection (*China Daily*, 12 November 1983). This represents about 0.5 per cent of the total industrial and agricultural output value of the nation, but it is significantly less than many other developed nations spend on environmental protection. Although there are about 20,000 persons working in the field of environmental protection, most of them were transferred from other departments and organizations, and only a small number have been specially trained in the field of pollution control (Shen, 1984). With most universities, which could produce trained scientists, technicians, and environmental managers, being closed during the 1966–76 Cultural Revolution, the pool of potential specialists trained in pollution control is of only limited extent.

China now accepts that it faces serious pollution problems and that rapid economic and industrial development may lead to pollution that could be harmful to national development in the long-term. What is not yet clear is the strength of commitment given to maintaining and improving air quality. China argues that economic development is not to be halted or

(accumulation in the bone) which results in tooth destruction, lameness and loss of weight. Whether fluoride air pollution can adversely affect human health is still debated but in areas with fluoride pollution from industry, children have been shown to have a decreased haemoglobin and increased erythrocyte level, in some cases with increased fluoride in their teeth, fingernails, hair and urine.

Figure 10.5 In many Chinese cities, as at Changchun in northeastern China, residential areas often contain industrial plants emitting black smoke
Source: C. R. L. Friend

slowed for the sake of environmental protection. With economic and industrial growth taking place so rapidly, with serious air-pollution problems having been inherited from previous decades, with limited availability of expertise in the field of pollution control, and with only the rudiments of a pollution-control equipment industry, there clearly remain doubts as to whether air quality can be safeguarded, let alone significantly improved, in the near future.

10.4 Socialism and Pollution

An examination of the pollution problems and pollution-control policies of the Soviet Union and China has highlighted the fact that pollution problems are as bad, and control policies as limited in their effectiveness, as in Western capitalist countries. In some ways, the situation in socialist countries is far worse, and this is particularly true of socialist countries in Eastern Europe. Carter (1985), for example, claims that Czechoslovakia is now among one of the most intensely air-polluted countries in the world. Even worse is the case of Poland, which, in the Katowice industrial region near the Czechoslovakian border, has the worst pollution in the world, according to Timberlake (1981). In 1983 it was announced that in this region, 430 per 100,000 inhabitants died prematurely because of environmental conditions, and that the death rate was 50 per cent above the national average. Circulatory diseases were 15 per cent more frequent, cancer 30 per cent and respiratory diseases 47 per cent more frequent. Infant mortality was 13 per cent higher.

Serious pollution occurs in socialist countries even though industry, the principal source of air pollution, is state-owned or state-controlled. It would seem relatively easy for the government to ensure that industries keep pollution to a minimum, but this clearly does not happen. There are several reasons for this situation, already mentioned in the foregoing case studies of the Soviet Union and China, but they need further elaboration in relation to all socialist countries. Of course, although abolishing private ownership of the means of production implies that a country is socialist, there are many forms of socialism, depending upon the Marxist interpretation adopted. Moreover, some researchers would argue that the existence of serious pollution problems reflects the fact that the state is deviating from the true socialist philosophy (Pepper, 1984).

Industrialization came decades, even centuries, later in socialist countries than in most Western nations. In parts of Eastern Europe industrialization got underway as late as the end of World War II. Many of the governments came to power with the avowed aim of modernizing their nation, supported by an ideology which promised to create the material abundance of capitalist countries. The national priority of modernizing the economy

Figure 10.6 Industrialists often find it cheaper to pay fines rather than install pollution-control equipment in Czechoslovakia
Source: R. A. Barnes

and improving the standard of living was interpreted as a need for rapid industrialization. Moreover, socialist countries reject the view that economic growth needs to be curbed so as to protect the environment. The traditional Marxist viewpoint plays down any concern for environmental degradation since it emphasizes the human ability to dominate and control nature. Like the developing nations, socialist countries are firmly committed to continued industrial expansion, and they hold that for all types of environmental damage, effective solutions already exist or can soon be invented. Protection of the environment has only been considered if it does not hinder the national drive for economic growth and a better standard of living. Pollution has not been ignored completely, but it has been tackled only when it was obvious or when it clearly indicated a wasteful use of natural resources.

Although concern for the environment was virtually absent in the early phase of industrialization, there has been a growing recent awareness that pollution can cause substantial economic losses to a nation. Consequently, the need for better pollution control has been expressed. However, the governments in socialist countries have found the decision to improve pollution control difficult to implement. This is because the pro-industry and pro-growth forces in the government hierarchy are well established. Production goals are still afforded greater importance than pollution control, with the latter frequently being regarded as non-productive in that it diverts resources away from increased production, whereas the act of pollution increases the economic growth and prosperity of the nation. Government desire for increased industrial productivity is reflected in the nature of environmental legislation. Many regulations are of a generalized nature, being difficult to interpret in practice. The strict pollution-control regulations which do exist are only loosely enforced. If fines are imposed on polluters, the fine is frequently small: for example, in Czechoslovakia, the payment of fines has proved much cheaper for enterprises than the cost of installing pollution-control equipment (figure 10.6). In Hungary, some industrial establishments incorporate fines into their budgets in advance rather than using less polluting technology (Laszlo, 1984). Surprisingly, the internalizing of the social costs of pollution in a production process is generally less advanced in socialist than it is in Western countries. The national emphasis on industrialization in some socialist countries is such that efforts are directed more towards the adaptation of the environment to industry. In Poland, for example, deciduous trees are being planted to replace sensitive Scotch pines, rather than industrial processes being modified in order to control the damaging sulphur dioxide emissions (Kormondy, 1980).

Even with rapid industrialization being pursued, socialist countries have had the choice of what particular aspects of industrialization to develop. Industries which minimized the amount and toxicity of wastes could have been emphasized. Socialist countries have long de-emphasized consumer goods and disposable products, private motor vehicles, and synthetics, and this has undoubtedly reduced the potential for pollution. However, since the genesis of socialist planning, all countries have emphasized heavy industry, which imposes the greatest burden on the environment. In more recent times many socialist countries have started to expand the production of consumer products and synthetic chemicals, thereby exposing themselves to the potential for forms of pollution which have long been part of Western economies. Some countries, being short of raw materials, have emphasized the importance of recycling wastes and this has reduced the potential for pollution. Unfortunately, because of the scarcity of fuel and primary raw materials, they have had to make maximum use of low-quality resources. Thus, for example, in a number of socialist countries, brown coal and lignite form the main fuels and these are low in calorific value and high in sulphur and ash content. Even when high-grade coal is available, as in Poland, the state of the economy requires that this coal be exported to ease the balance-of-payments problem.

Socialist planning should prevent excessive regional concentration of production and wealth by creating a more even distribution of population, cities and industry. Such a situation should result in a dispersal of pollution and fewer areas of concentrated environmental degradation. However, while a conscious – albeit selective – geographic dispersal into thinly populated but resource-rich areas has unquestionably been part of the Soviet experience, industrial decentralization is much less evident in other socialist countries. In the beginning of the socialist era, the industrial transformation of backward areas and small urban centres was attempted, but the desire for rapid industrial growth gave priority to the expansion of manufacturing in large cities and established industrial regions (Dienes, 1974). Even China, with its much publicized societal decentralization, retains its heavy industry in selected regions, with the consequence that serious regional pollution occurs.

Most socialist countries lack effective environmental pressure-groups. The various media are state-controlled and information which points to inadequacies in environmental protection may be suppressed. The public is often unaware of the true state of the environment. All of these factors reduce the pressure on the central government to ensure effective pollution control. Just what can occur in a socialist country when freedom of access to information is gained was demonstrated during the short-lived rise of the Solidarity trade-union movement in Poland in the early 1980s: the appalling state of the environment emerged as an issue and the worst polluters were closed down. However, it cannot be assumed that environmental protection will always gain the support of the public. Public pressure for a higher standard of living, especially reflected in better quality and greater availability of consumer goods, may mean that environmental protection remains subservient to this goal.

One may conclude that Socialist countries seem as prone to pollution problems as Western nations. As an indicator of a nation's potential for pollution or the willingness and effectiveness with which it will deal with pollution problems, it seems to make little difference whether the economic system is socialist or capitalist. The replacement of private greed for profits with public greed for prosperity seems to make little difference to air-quality levels. Both forces appear to play much the same role in fostering industrial development, which is the primary cause of pollution problems, and in obstructing policies and programmes to safeguard the environment.

11 International Collaboration on Pollution Control

11.1 Current and Potential International Pollution Problems

The pollution-control policies developed initially by industrialized nations were concerned with tackling air-pollution problems caused by emission sources within their own countries. National policies gave no consideration to pollution being exported or imported. In order to solve local pollution problems, national policies often encouraged or even required industry to adopt taller industrial stacks so as to disperse pollutants away from the immediate vicinity: no consideration was given to the possibility that these pollutants might be transported over long distances and cause adverse effects in other countries. National policies also failed to recognize that pollutants might cause adverse environmental effects at a global scale. It was only during the 1970s, when an increasing number of transfrontier and global pollution problems and threats emerged, that nations began to direct serious attention towards the need to control pollutants being released beyond their jurisdiction.

Transfrontier pollution problems include acid rain, photochemical oxidant (ozone) episodes, and accidental releases of large quantities of pollutants such as ionizing radiation and toxic chemicals. Global problems range from the build-up in levels of carbon dioxide, toxic chemicals, ionizing radiation and anthropogenic heat, to the depletion of stratospheric ozone. Solutions to such transfrontier and global pollution problems will emerge only when nations are willing to co-operate to undertake effective pollution-control measures. Such international collaboration is encouraged by a number of important international organizations. This chapter assesses the international progress made in tackling the problems of acid rain, stratospheric ozone depletion and the carbon dioxide build-up.

11.2 International Environmental Law

International environmental law, which could be used to address trans-frontier and global pollution problems, is only in an embryonic stage of development, being comprised mainly of doctrines set forth in multinational agreements, declarations of international organizations, and a few international tribunal decisions. The clearest statement of international law as it relates to air pollution is provided by an international arbitration in the 1930s between Canada and the United States. Pollution from a smelter at Trail, British Columbia, was causing damage in adjacent areas of the state of Washington. The decision in the Trail Smelter Arbitration reads . . .

> under the principles of international law, as well as the law of the United States, no state has the right to use or permit the use of its territory in such a manner as to cause injury by fumes in or to the territory of another or the properties or persons therein, when the case is of serious consequence and the injury is established by clear and convincing evidence.

The tribunal required both payment of damages and the establishment of a programme to monitor and reduce pollution from the smelter.

The decision of the Trail Smelter tribunal clearly influenced the wording of the Declaration on the Human Environment agreed at the United Nations Conference at Stockholm in 1972. This conference, attended by 114 nations and 37 intergovernmental organizations, is credited with marking the beginning of an awareness among national governments that many environmental problems can be solved only by international collaboration. The Declaration contained two important principles for environmental protection: firstly, Principle 21 provides that:

> states have, in accordance with the Charter of the United Nations and the principles of international law, the sovereign right to exploit their own resources pursuant to their own environmental policies, and the responsibility to ensure that activities within their jurisdiction or control do not cause damage to the environment of other states or of areas beyond the limits of national jurisdiction.

Secondly, Principle 22 provides that

> states shall co-operate to develop further the international laws regarding liability and compensation for the victims of pollution and other environmental damage caused by activities within the jurisdiction or control of such states to areas beyond their jurisdiction.

While such principles constitute important first steps towards providing a comprehensive detailed framework of international environmental law, they are of limited use in practice. Thus, for example, no guidance is given in Principle 21 for deciding at what point a nation's interest in pursuing industrial development is outweighed by concerns surrounding the environmental effects of transfrontier pollution. Another problem in tackling transfrontier pollution problems is the need to prove that pollution damage in one country is due to emissions from specific sources in another country. Except in exceptional situations, such as in the case of the Trail smelter, it is extremely difficult to prove: when pollutants are transported over very long distances, obtaining proof that specific sources were the cause of pollution damage in another country is virtually impossible. Even if such evidence were available and compensation for the pollution damage were to be claimed by the affected country, there remain the problems of determining the appropriate amount and enforcing payment of the compensation. Finally, a discussion of the problems of the difficulties of assessing and arranging compensation assumes that victims of pollution would want such a scheme to operate. It may be that countries suffering from foreign pollution have fundamental objections to the introduction of compensation schemes. Norway, for example, believes that a compensation scheme for the effects of acid rain trivializes the problem and implicitly sanctions continued transboundary pollution. They argue that a polluter would always prefer to pay what would be token compensation rather than reduce emissions (Wetstone and Rosencranz, 1983). It is with this in mind that Scandinavian countries have rejected any financial assistance for their remedial programme involving the liming of their lakes.

To date, it appears that the international legal structure offers useful principles of environmental responsibility, but they are neither sufficiently defined nor sufficiently enforceable to support effective application to specific controversies.

11.3   Acid Rain

11.3.1   Introduction

The problem of acid rain (acid deposition) has received increasing international attention in recent years. Whereas at the time of the Stockholm Conference on the Human Environment, acid rain affected only a few Western industrialized countries, it is now recognized as a problem facing most industrialized countries as well as an increasing number of developing nations such as Brazil, China and India. It is a particularly sensitive international issue because countries with the greatest emissions of the precursor pollutants which produce acid rain are not those suffering the greatest damage. Thus, Scandinavia experiences acid rain effects as a result of pollution emissions in other European countries, while Canada suffers from emissions originating in the industrial regions of the United States. However, all industrialized countries export and import pollution, albeit in differing relative proportions, and this makes the problem of acid rain both very complex and a truly international issue.

Although there are shortcomings in our understandings of the causes and effects of acid rain, it is generally agreed there is sufficient knowledge to justify taking some immediate action to reduce the effects of acid rain. Yet having made the decision to act, there are many questions which must then be addressed: which pollutants should be tackled and from which sources, what amounts of reduction are needed, and in which countries should this action be taken? Sulphur dioxide and oxides of nitrogen are the most obvious pollutants for which emission reduction should be sought, but there is increasing evidence that photochemical oxidants make a significant contribution to 'acid rain' damage done to crops and forests. Having decided which pollutants to tackle, one must decide whether specific emission sources should be tackled, or whether an overall reduction in emissions from all sources should be considered. Although there is general agreement on tackling those large emission sources such as power stations with high stacks, emissions of oxides of nitrogen from motor vehicles also need to be reduced. The technical feasibility of ways to reduce emissions from stationary or mobile sources of these pollutants must then be assessed.

The weighing-up of the costs of the environmental benefits to be gained from a reduction in acid rain damage against the economic and social cost of an emission reduction programme provides the answer to the question of what quantity of emission reduction can in practice be achieved. In this connection, uncertainty exists as to whether there is a linear relationship between emissions of sulphur dioxide and oxides of nitrogen, and the resulting acid deposition. Some researchers suggest that the hydrocarbon and photochemical oxidant concentrations are the main determinants of the rate of conversion of sulphur dioxide and oxides of nitrogen to acids in the atmosphere. If this is so, then the reduction of emissions may not necessarily reduce acid deposition. Data collected in Europe points to a non-linear relationship in that there is not enough available oxidant in the atmosphere to acidify all of the emitted sulphur dioxide and oxides of nitrogen. To cut emissions by 50 per cent would, on this evidence, be very unlikely to produce a corresponding reduction in acid rain. In the United States, the US National Research Council (1983a) has concluded that there is a proportional relationship between emissions of sulphur dioxide and oxides of nitrogen and acid deposition over the eastern half of the United States, when large areas and time-scales are considered. In other words, a 50 per cent reduction in emission of these gases would yield a 50 per cent reduction in acid deposition. However, some researchers express reservations concerning this conclusion (Hidy et al., 1984).

Given the costs involved in reducing emissions, agreement as to which countries should reduce their emissions by, say, between 30 and 50 per cent, is fraught with difficulties. If one applied the Polluter Pays Principle, then it would be those countries exporting more pollution than they import

Figure 11.1 Liming is used to counteract the effects of acidification in Swedish lakes, but it is a temporary measure only
Source: R. A. Barnes

that should be required to reduce emissions. However, given that the introduction of pollution-control measures would increase the price of industrial production and affect international trading, many countries would be reluctant to act if others did not do so. Consequently, action at present and in the near future is likely to be limited to obtaining agreement that all countries reduce their emissions by a similar proportion.

11.3.2   Options Available for Tackling Acid Rain

Although energy conservation and the use of nuclear power and/or renewable energy sources are means of reducing emissions of sulphur dioxide and oxides of nitrogen, these are long-term solutions only. The seriousness of the acid rain problem at the present time requires solutions which can significantly reduce acid rain effects within the next ten years.

11.3.2.1   Liming of lakes and streams   A widely-used method of counteracting the effects of lake acidification where poor natural buffering exists is to add calcium, usually in the form of lime, to the water (figure 11.1). This in turn raises the pH level of the water and reduces the amount of aluminium in solution. The liming of about 3000 lakes in Sweden has enabled some lakes to be restocked and has resulted in the recovery of some fisheries (McCormick, 1985; UK House of Commons Environment Committee, 1984). Liming of all 'at risk' lakes in Norway and Sweden would amount to about £60–100 million per annum (Environmental Resources Limited, 1983).

Liming has some practical disadvantages in that it is difficult to apply to lakes in remote locations and in streams. Although liming leads to some immediate improvements in fish stocks, it has ecological disadvantages. Important fish-food types can be lost or reduced. Added lime cannot act quickly enough to neutralize the rapid introduction of acids and heavy metals produced during spring snowmelt. The addition of calcium to a lake does not prevent the leaching of toxic aluminium from soils, but merely causes it to precipitate to the bottom of the lake. If the liming is stopped then a further fall of acid rain brings the aluminium back into solution and it would then kill off the fish with which the lake has been restocked. Liming appears to be only a temporary measure, until significant reductions in emissions of the precursors of acid rain are achieved. Liming has been shown to be ineffective in protecting terrestrial ecosystems and Swedish experiments discovered that forest growth is retarded by the addition of lime (Swedish Ministry of Agriculture, 1982).

11.3.2.2   Using fuels with a lower sulphur content   The sulphur content of fossil fuels can range from 0.5 per cent to over 5 per cent. Although simply switching from higher-sulphur to lower-sulphur fuels is one easy

way of achieving emission reductions, and is the method usually preferred by industry, low-sulphur fuels are limited in availability on the international market. Thus, for example, most North American low-sulphur oil supplies are used domestically; North Sea supplies are finite and are currently at their peak production level; and the remaining supplies (Algeria, Libya, and Nigeria) are currently subject to OPEC production ceilings. Supplies of low-sulphur coal (with less than 1 per cent sulphur content) are also restricted, and even in the United States, where coal reserves are sufficient for fuel switching to take place, this option raises many important questions concerning regional economic impacts, employment in the coal industry, and the ability of the coal industry to accommodate a rapid shift from high- to low-sulphur coal production. Coal production would have to shift away from the higher-sulphur supply regions (such as North Appalachia and the Midwest) and toward lower-sulphur supply regions (such as Central Appalachia and the Rockies).

Fuel oil can be desulphurized, although at a cost which rises very sharply with the volume supplied and the degree of sulphur removal required. Although a 30 per cent reduction in average sulphur content can be achieved, any further reduction is technically difficult and it is highly questionable whether it would be economically feasible (Environmental Resources Limited, 1983). In the case of coal, physical cleaning or washing is one well-established means of producing a fuel with a lower sulphur content. Sulphur in coal occurs in two main forms: organic and pyritic. Organic sulphur, which is an integral part of the coal matrix and which generally cannot be removed by direct physical separation, comprises from between 30 and 70 per cent of the total sulphur content of most coals. Pyritic sulphur (iron sulphide) is the mineral pyrite which occurs in coal as discrete and sometimes microscopic particles. Given that it is a heavy mineral with a specific gravity of about 5.0, as compared with a maximum specific gravity of 1.8 in coal, the pyritic content of coal can be significantly reduced by crushing and subsequent specific-gravity separation (Economic Commission for Europe, 1985). Although the pyritic sulphur content of coals varies, up to about 40 per cent of the sulphur content of coal could be removed in this way (Barnes, 1986).

The cleaning of coal is already common practice in many countries. The reason for cleaning is not in order to reduce sulphur dioxide emissions when the coal is burned, but rather, to remove mineral impurities and so improve the coal's quality. For example, two-thirds of coal is already washed in the United Kingdom (removing about 15 per cent of the pyritic sulphur and leaving a coal with an average sulphur content of 1.6 per cent): the remainder is considered too fine to be washed economically, and is therefore blended in a mix with washed coal. This situation means that the potential for further reduction of the sulphur content of coal in the United Kingdom is limited and the UK House of Commons Environment

Committee (1984) suggests that further cleaning of power-station coal would remove only another 6 per cent. In contrast, conventional washing of all coal from eight states in the eastern United States would achieve an additional reduction in sulphur dioxide emissions of over two million tonnes. More intensive cleaning (crushing to a smaller size and separating at a lower specific gravity) would raise the estimated sulphur dioxide reduction to over three million tonnes (Green, 1984).

Although fuel switching and coal cleaning are two proven low-cost options for achieving a reduction in sulphur dioxide emissions, the total amount of emission reduction is insufficient to meet national emission-reduction targets of, say, 50 per cent (equivalent in the United States to about 12 million tonnes) within ten years.

11.3.2.3   Fluidized bed combustion systems   A fluidized bed can be described as a bed of finely divided particles through which a gas is forced from below. At a critical gas velocity the bed becomes turbulent with the suspended particles mixing rapidly, behaving like a fluid in motion. In a coal-fired fluidized bed, finely divided coal is burned in a granular bed of mineral material, generally limestone or dolomite, together with residual ash from previously burned coal. The fuel can be fed continuously to the bed while sufficient air for combustion is forced through. Most of the sulphur dioxide reacts with the bed material to form calcium sulphate, which can then be withdrawn from the system with the ash. Fluidized Bed Combustion (FBC) systems can operate either at atmospheric pressure (atmospheric FBC) or at pressures up to ten atmospheres or higher (pressurized FBC).

Sulphur dioxide emissions can be reduced by 90 per cent and the low temperature (about 900°C compared with 1650°C in conventional coal-fired operations) of the combustion keeps down the formation of oxides of nitrogen (a reduction of 30–40 per cent). FBC can be used efficiently with almost any fuel, including municipal wastes, paint residues, sewage sludge and oily wastes. A multi-fuel FBC system permits switching to solid, liquid, or gaseous fuels so that changes in the cost and/or availability of different fuels could be taken advantage of. Given that FBC improves the efficiency of the combustion system, then it provides an economic gain over alternative combustion systems. The major current limitation of FBC is that technical difficulties have limited commercially viable units constructed to date to 150 MW or less (Economic Commission for Europe, 1985).

11.3.2.4   Flue gas desulphurization systems   Several technologies have been developed for reducing emissions of sulphur dioxide generated during the combustion process. Flue Gas Desulphurization (FGD) systems (or 'scrubbers') vary according to the type of material used for sulphur dioxide absorption, the nature of the by-products, and the ability of the process to remove other pollutants, in particular oxides of nitrogen. The most common technology presently in use is the wet scrubbing process based on spraying a slurry of lime or limestone into the flue gas. The principal advantage of lime over limestone is the higher sulphur dioxide removal efficiencies (the system is usually designed to achieve up to 90 per cent reduction, but even a 99 per cent reduction is possible). These processes can be either throw-away or by-product recovery processes. In the United States, it is wide-spread practice in wet scrubbing to produce an unusable sludge which must be disposed of, for instance, in a settling pond. In this way, the wastes are merely being transferred from one part of the environment to another, even though this method may be the 'best available environment option'. In Europe, the processes of interest are those which produce a usable by-product, usually gypsum, for use in the cement and construction industries.

Whichever FGD system is preferred, the capital investment is typically 20 per cent or more of the total cost of, say, a coal-fired power plant. Most pollution-control programmes, formulated as they are to achieve substantial reductions of sulphur dioxide emissions within, say, ten years, inevitably rely on retrofitting FGD systems to major emission sources such as coal-fired power stations.

11.3.2.5   Reducing oxides of nitrogen emissions from stationary sources   During combustion, oxides of nitrogen may be formed by oxidation of molecular nitrogen from the combustion air (thermal oxides of nitrogen), and by oxidation of the nitrogen which is chemically bound in the fuel and is released during combustion (fuel oxides of nitrogen). The amount of thermal oxides of nitrogen produced is primarily a function of flame temperature, whereas the amount of fuel oxides of nitrogen produced depends mainly upon the thermal and chemical fate of the coal particles in the flame. In a typical boiler, burning a standard coal with a nitrogen content of 1 per cent, the fuel oxides of nitrogen can represent about 75 per cent of the total oxides of nitrogen (Economic Commission for Europe, 1985). The formation of oxides of nitrogen is strongly influenced by the residence time of the fuel in the hottest part of the flame. By lowering the temperature in the hot part of the flame or by reducing the quantity of oxygen in the centre of the flame and shortening the throughput time of the fuel, significant reductions in emissions can be achieved (Swedish Ministry of Agriculture, 1982). Burners redesigned in this way are usually called 'low-oxides of nitrogen burners' and are capable of reducing the emission of oxides of nitrogen by 30–40 per cent, or even 60 per cent. The cost of retrofitting low-oxides of nitrogen burners to a 2000 MW coal-fired power station would be about £10 million. In addition, flue gas denitrification systems are being developed which can achieve reductions of 80–90 per cent.

11.3.2.6   Reducing emissions of oxides of nitrogen and hydrocarbons from motor vehicles   Oxides of nitrogen from motor vehicles not only add to the acid rain problem, but along with hydrocarbons, they also lead to the production of photochemical oxidants (ozone) which are believed to make a significant contribution to 'acid rain' damage of crops and forests in parts of central Europe and the United States. Generally, motor vehicle emissions of oxides of nitrogen contribute 7–10 per cent of the total acidity in wet and dry deposition in Europe. There are several ways in which emissions of oxides of nitrogen (and hydrocarbons) can be reduced from motor vehicles. The United States and Japan already employ three-way catalytic converters which operate using unleaded petrol and remove 97 per cent of oxides of nitrogen and 90 per cent of hydrocarbon emissions (refer to chapter 8). Lean-burn engines are the alternatives to catalysts. In these engines, total fuel combustion is virtually attained by mixing with relatively large quantities of turbulent air. Lean-burn engines have the additional advantage of fuel economy improvements over and above conventional engines of up to 15 per cent, in contrast with the use of catalytic converters which can increase fuel consumption. Such engines generally add less to the cost of a motor vehicle than catalytic converters.

Engines produce more oxides of nitrogen as speed and engine heat increase. Traffic densities in central Europe are particularly high and where lax speed restrictions apply, large emissions of oxides can be produced. The lowering and stricter enforcement of speed limits could produce reductions in oxides of nitrogen with little expenditure. However, the significance of speed restrictions for the acid rain problem is questioned: for example, a government inquiry in West Germany in 1985 concluded that high-speed driving contributed about 10 per cent to the total emissions of oxides of nitrogen from motor vehicles. As these emissions themselves contributed only 10 per cent to acid rain, speed restrictions would likely result in a reduction in acid rain effects of at most 1 per cent in West Germany.

*11.3.3   Progress Towards a United Nations Convention
         on Acid Rain*

Following the United Nations Conference on the Human Environment in 1972, which first alerted the world to the worsening of the acid rain problem in Europe, international attention on acid rain was maintained by the Organization for Economic Co-operation and Development (OECD). This organization was founded in 1960 to promote sound economic growth and greater international trade and in this context it addressed international environmental issues. Members include the countries of Western Europe, North America, Australasia, and Japan.

The OECD provided the first international co-operative assessment of the long-range transport of sulphur pollutants in Western Europe by initiating The Long-Range Transport of Air Pollutants (LRTAP) study in 1972. Using 1974 data from 11 European countries, this project concluded that although the countries with the largest sulphur dioxide emissions also received the largest acid deposition, five Western European nations, Norway, Sweden, Finland, Austria, and Switzerland, imported more pollution from other countries than they received from their own sources (OECD, 1977). The OECD urged member countries to reduce pollution emissions and pressed for the Polluter Pays Principle (of internalizing rather than externalizing the costs of pollution) to be applied internationally to avoid distortions in international trade and investment which might put industries or countries with stringent pollution-control policies at a competitive disadvantage.

The OECD study was succeeded in 1977 by The Co-operative Programme for Monitoring and Evaluation of Long-Range Transmission of Air Pollutants in Europe (EMEP), organized under the auspices of the United Nations Economic Commission for Europe (ECE) and with financial support from the United Nations Environment Programme (UNEP). Using data from 70 monitoring stations in 20 countries, EMEP attempted to quantify what proportion of sulphur pollutants originates from outside a country's territory (table 11.1). Assuming that foreign sources are responsible for half of the deposition classified as of 'undecided' origin, this study highlights the fact that foreign pollutants account for about 77 per cent of the sulphur deposition in Norway, 70 per cent in Sweden, 64 per cent in Finland, 80 per cent in Austria, and 84 per cent in Switzerland.

The countries with the largest emissions of sulphur dioxide include the USSR, the United Kingdom, East Germany, West Germany, France, Italy, Czechoslovakia, Poland and Yugoslavia (table 11.2). However, the relative importance of each country as an exporter of pollution depends upon its geographical position and meteorological conditions. Thus, for example, given the high frequency of winds from the west, the United Kingdom is in a prime position to export pollution to the rest of Europe. On the other hand, during anticyclonic conditions, when short-term episodes of very high pollution concentrations occur throughout Europe, it is emissions originating in Czechozlovakia, East Germany, West Germany and Poland, transported by easterly winds, which are of most importance in Western Europe.

Estimates of the economic costs of acid rain in Europe remain uncertain but the OECD (1980a) estimates that whereas member countries spend between 1 and 2 per cent of their Gross National Products on pollution abatement, the damage from pollution amounts to between 3 and 5 per cent of GNP. One assessment of the annual costs of acid rain for the European

Table 11.1 EMEP-calculated sulphur deposition due to foreign emissions, the undecided background deposition and the deposition from indigenous sources, as a percentage of total deposition in each European country and based on average monthly data for the period October 1978 to September 1980

Country	Deposition from foreign sources (% of total deposition)	Undecided background deposition (% of total deposition)	Deposition from indigeneous sources (% of total deposition)
Albania	67	18	15
Austria	76	9	15
Belgium	53	6	41
Bulgaria	47	8	45
Czechoslovakia	56	7	37
Denmark	54	10	36
Finland	55	19	26
France	34	14	52
German Democratic Republic (East Germany)	32	3	65
Germany, Federal Republic of (West Germany)	45	7	48
Greece	51	12	37
Hungary	54	4	42
Iceland	25	75	0
Ireland	32	40	28
Italy	22	8	70
Luxembourg	73	0	27
Netherlands	71	6	23
Norway	63	29	8
Poland	52	6	42
Portugal	33	40	27
Romania	56	8	36
Spain	18	19	63
Sweden	58	24	18
Switzerland	78	12	10
Turkey	39	19	42
USSR	32	15	53
UK	12	9	79
Yugoslavia	41	8	51

Source: Anonymous, 1982

Community lists a loss of revenue from tourism and sport fishing of £15 million, a 5 per cent loss of value of German forests together with further limited damage elsewhere of £300 million, a loss of crop value as a result of the effects of ozone, sulphur dioxide and oxides of nitrogen of £900–1100 million, building damage costs of £540–2700 million, and costs of drinking water treatment in some areas of £20 million (Environmental Resources Limited, 1983). The OECD tentatively estimated in 1982 that the benefits in the OECD European countries as

Table 11.2   European emissions of sulphur dioxide and oxides of nitrogen for 1978 and 1980

	SULPHUR DIOXIDE (million tonnes)		OXIDES OF NITROGEN (million tonnes)	
	1978	1980	1978	1980
Albania	0.100	—	—	—
Austria	0.430	0.444	0.547	—
Belgium	0.808	0.809	—	0.150
Bulgaria	1.000	1.000	—	—
Czechoslovakia	2.950	3.100	—	—
Denmark	0.455	0.455	—	0.241
Finland	0.560	0.570	0.178	—
France	3.385	3.262	1.430	1.455
Germany East	4.000	—	—	—
Germany West	3.550	3.580	3.000	—
Greece	0.700	0.700	0.196	—
Hungary	1.640	1.633	—	—
Ireland	0.250	—	—	—
Italy	4.400	3.800	1.270	—
Luxembourg	0.030	—	—	—
Netherlands	0.490	0.440	0.505	0.506
Norway	0.150	0.143	—	0.073
Poland	2.500	2.755	—	0.247
Portugal	0.187	0.264	0.196	—
Romania	0.200	0.200	—	—
Spain	3.570	—	0.772	—
Sweden	0.530	0.510	0.320	0.320
Switzerland	0.124	0.119	0.161	—
Turkey	1.000	1.000	0.380	—
USSR	24.500	25.000	—	—
UK	5.02	4.67	1.796	1.785
Yugoslavia	2.950	3.000	—	—

Source: Information is derived from OECD sources

a whole lay in the range £80 to £800 per tonne of sulphur dioxide removed, based on a consideration of the economic value of health effects, corrosion of materials, crop damage and acidification of waters.

Towards the end of the 1970s, the OECD involvement in the acid rain problem receded somewhat as the United Nations Economic Commission for Europe (ECE) initiated a major step forward in international collaboration. In 1979, a conference of the 34 member countries of the ECE (including the Eastern European nations) formally adopted the Convention and Resolution on Long-range Transboundary Air Pollution (LRTBAP) and this Convention officially came into force in 1983 (table 11.3). The Convention commits signatories to the attempt of limiting and, as far as is possible, gradually reducing and preventing air pollution, including long-range transboundary air pollution; and that, in order to

Table 11.3 Signatories to the Convention on Long-range Transboundary Air Pollution and details of those countries who have joined the Thirty-Per-Cent Club

Signatory	Date ratification	Date of accession to the Thirty-Per-Cent Club	Promised reductions of SO_2, from 1980
Austria	Dec. 1982	June 1983	50% by 1995
Belgium	July 1982	June 1984	50% by 1995
Bulgaria	June 1981	June 1984	30% by 1993
Byelorussian SSR	June 1980	June 1984	30% by 1993
Canada	Dec. 1981	June 1983	50% by 1994
Czechoslovakia	Dec. 1983	Sept. 1984	30% by 1993
Denmark	June 1982	June 1983	50% by 1995
Finland	April 1981	June 1983	50% by 1995
France	Nov. 1981	March 1984	50% by 1990
Germany (Democratic Republic)	June 1982	June 1984	30% by 1993
Germany (Federal Republic)	July 1982	June 1983	60% by 1993
Greece	Aug. 1983		
Hungary	Sept. 1980	April 1985	30% by 1993
Iceland	May 1983		
Ireland	July 1982		
Italy	July 1982	Sept. 1984	30% by 1993
Liechtenstein	Nov. 1983	June 1984	30% by 1993
Luxembourg	July 1982	June 1984	30% by 1993
Netherlands	July 1982	March 1984	40% by 1995
Norway	Feb. 1981	June 1983	50% by 1994
Poland	March 1985		
Portugal	Sept. 1980		
Spain	June 1982		
Sweden	Feb. 1981	June 1983	65% by 1995
Switzerland	May 1983	June 1983	30% by 1995
Turkey	April 1983		
Ukrainian SSR	June 1980	June 1984	30% by 1993
USSR	May 1980	June 1984	30% by 1993
United Kingdom	July 1982		
USA	Nov 1981		
European Community	July 1982		

Source: Compiled from information provided by UNEP

to achieve this, they shall use the best available technology that is economically feasible. Specifically, the Convention provides for exchange of information, monitoring (improvements in EMEP allowing modelling of sulphur deposition throughout Europe as illustrated in figure 11.2) and research as a basis for the development of strategies and policies to combat air pollution (Kelley, 1980). Its significance is enormous, given that this was the first time that nations of Eastern Europe had joined with Western

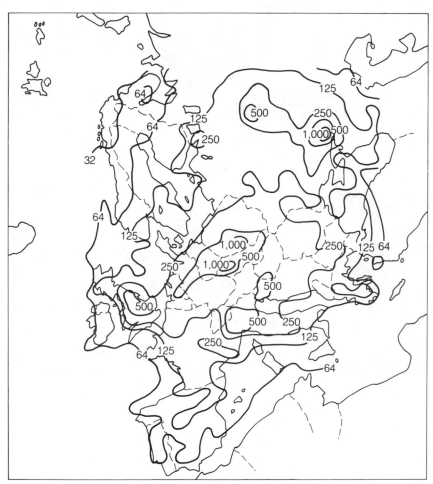

Figure 11.2   EMEP-calculated average monthly dry plus wet deposition of sulphur in Europe from all emissions for the period October 1978 to September 1980
Units: mg.m^2 as sulphur
Source: Eliassen and Saltbones, 1983

European and North American nations in an environmental agreement. The Convention enjoyed a fortuitous advantage as a pawn in East-West relations at a time when the Soviet leader, Leonid Brezhnev, was seeking an area for international collaboration. Had the signing of the Convention been scheduled two months later, after the intervention of the Soviet Union in Afghanistan, there would probably have been no accord (Wetstone and Rosencranz, 1983). Although the Convention is a major step forward in international environmental collaboration, it lacks any specific targets or timetable for the reduction in emissions which are ultimately needed to reduce the effects of acid rain. Scandinavian countries urged that a programme of specific emission reductions be made part of the Convention, but this was opposed by several countries, most notably the United States, the United Kingdom, and West Germany.

The Stockholm Conference on Acidification of the Environment in 1982 provided the opportunity for several countries to restate the need for effective international collaboration in reducing sulphur dioxide emissions. This Conference produced the unexpected about-face of West Germany. In 1979, West Germany had opposed any form of the LRTBAP Convention which would commit it to specified emission reductions, but in 1982, following the discovery of extensive and worsening forest die-back, West Germany called for substantial emission reductions from ECE member countries. To some extent, West Germany became more effective than the Scandinavian countries in urging for emission controls since West Germany is itself a major source of emissions of sulphur dioxide and oxides of nitrogen. Subsequently, at a meeting of the Convention's Executive Body in June 1983, a number of countries, including the Scandinavian countries, West Germany and Canada, committed themselves to reducing sulphur dioxide emissions by 30 per cent by 1993, based on 1980 levels. Some countries even made commitments to reduce emissions by 50 or 60 per cent by 1994 or 1995 (table 11.3).

Eastern European countries agreed to cut 'exported pollution' by 30 per cent, although Poland, with its severe economic problems, did not sign the agreement. The emphasis placed by Eastern European countries on 'exported pollution' reflects their view that only pollutants carried to other countries should be the subject of international co-operation. What happened inside a country, they argued, was strictly an internal matter. With so many nations willing to join the 'Thirty-Per-Cent Club', this emission-reduction commitment was adopted as a protocol to the LRTBAP Convention in 1985 and it entered force in October of that year (McCormick, 1985). Many researchers would argue that the 30 per cent reduction is only a precautionary measure and that if the problem of acid deposition is ultimately to be solved, reductions of 75–80 per cent in both sulphur dioxide and oxides of nitrogen emissions may be needed.

The two most important nations not to have joined the Thirty-Per-Cent Club (or to have signed the equivalent Convention protocol) are the United Kingdom and the United States (refer to section 11.3.5). The United Kingdom government made it clear that it was 'not prepared to commit this country to expensive methods of abating sulphur dioxide emissions when the benefits are so uncertain' (UK Department of Environment, 1984b). However, the United Kingdom has declared that it shares the objectives of these countries and that it intends to achieve a 30 per cent reduction from the 1980 level, though not until the late 1990s. In other words the United Kingdom offers only an intention to reduce its emissions, rather than a formal commitment, and the 30 per cent reduction will take longer than the timetable being followed by members of the Thirty-Per-Cent Club. One reason for the United Kingdom's reluctance to commit itself formally is that by 1985, sulphur dioxide emissions had already decreased by 42 per cent since 1970, and by 24 per cent since 1980. It believes the choice of 1980 as the reference level to be unfair, given that emissions in the United Kingdom peaked much earlier than in other European nations. However, the reduction in emissions was not entirely due to pollution-control policies, but rather, partly to the economic recession, partly to the change in the nature of industries, and partly to a switch from fuel oil or coal to natural gas for private industrial and domestic use. Looking ahead, the government of the United Kingdom believes that its plans for increased reliance on nuclear power to generate electricity would bring about further reductions in sulphur dioxide emissions without having either to strengthen its pollution-control policies or to invest vast sums on additional pollution-control equipment, even though this further reduction in sulphur dioxide emissions will not occur until the late 1990s.

International pressure on the United Kingdom to reduce its emissions by 30 per cent by 1993 will intensify, given that the United Kingdom is the largest producer of sulphur dioxide in Western Europe in the 1980s. European countries are most concerned about emissions from Britain's large (2000 MW) coal-fired power stations whose tall stacks, used for dispersal of sulphur dioxide emissions, are condemned as encouraging long-range transport of pollutants (figure 11.3). Two-thirds of the nation's sulphur dioxide emission comes from these power stations and Britain is singled out as the country which is the highest contributor to sulphur deposition in Scandinavia. How long the United Kingdom can resist international pressure to agree formally to a 30 per cent emission reduction is unclear, but two developments make it likely that it will not be for long. Firstly, acid rain effects are increasingly being observed within the United Kingdom and this may exert the necessary internal public and political pressure, as it did in West Germany, officially to sign the LRTBAP Convention protocol and join the Thirty-Per-Cent Club. Secondly, the United Kingdom is a member of the European Community and the viewpoint of most other Member States is that a substantial emission reduction programme is needed and that Britain must abide by the majority viewpoint.

11.3.4   The European Community Response

Although acid rain affects all countries in Europe, the group of countries able to take the most effective action to reduce acid deposition belong to

Figure 11.3 Environmental groups singled out United Kingdom power station emissions in their campaign to tackle acid rain in Europe. This postcard, sent to British hotel-keepers and tourist boards in 1985–6, was part of an international tourist boycott by over two million people to protest about the United Kingdom's reluctance to join the Thirty-Per-Cent Club
Source: Friends of the Earth International

the European Community (EC), since it is the only international organization able to pass international pollution control policies which are binding on Member States. In the ECE, the Eastern European countries are likely to resist adoption of strong control measures, and so many European countries, such as Scandinavia and Switzerland (with its increasing forest die-back problem), look to the EC to begin the process of making major emission reductions in Europe. Community agreement is seen as an essential step to achieving subsequent agreement within the broader group of countries belonging to the ECE.

Action by the EC in reducing emissions of sulphur dioxide and oxides of nitrogen has already been taken in several forms. Within the framework of the Community's policy on environmental protection, laid down in the successive 'Action Programmes on the Environment', various pollution control Directives have been adopted. These refer to the establishment of air-quality standards, the limiting of the amount of sulphur in gas-oil, and the requirement that industrial plants over 50 MW, including power stations, must receive prior authorization. The latter requirement, agreed in 1984 and effective from 1987, requires that the authorities be satisfied that a new plant will not give rise to significant air pollution and that appropriate preventative measures, including the application of the 'best available abatement technology, provided it does not give rise to excessive costs', will be taken into account (this endorses the traditional United Kingdom concept of 'best practicable means'). This Directive also includes provision for the Council, if necessary, to fix Community-wide emission limits.

The EC Directive which would result in the most significant reduction in pollution emissions is the current Draft Directive on the limitation of emission of pollutants from large combustion plants. The Directive limits the total emission from combustion plants over 50 MW in each country and requires, by 1995, an emission reduction of 60 per cent in sulphur dioxide, 40 per cent in oxides of nitrogen and 40 per cent of particulates, taking 1980 as the base year. Complying with the sulphur dioxide emission reduction poses the greatest difficulty and cost for Member States. It would be up to the Member States to choose whether to load all the emission reductions on to one type of plant, such as power stations, or to distribute them more evenly. The draft proposal also requires that emissions from new plant, or those having undergone substantial alter-ation, comply with stringent emission limits, although it is envisaged that for plant less than 100 MW compliance will be deferred until 1990. The standards are tighter for plant greater than 300 MW such that FGD would probably be necessary, while plant below 300 MW would probably require the use of low-sulphur fuel (coal or oil less than 1 per cent sulphur).

In 1980, total UK sulphur dioxide emissions amounted to about 4.7 million tonnes (table 11.2), of which 3.6 million tonnes were from about 450 plants greater than 50 MW, with 2.9 million of these from coal-fired power stations. The proposed EC Directive would reduce sulphur dioxide emissions by around 2.2 million tonnes, or 47 per cent. Obviously this would place the United Kingdom (and other Member States) in the Thirty-Per-Cent Club. Given the dominance of emissions from power stations, this reduction could be achieved by considering only the major Central Electricity Generatoring Board (CEGB) coal-fired power stations, with fourteen being over 1000 MW. To meet the deadline for the EC Directive – that is, to achieve a 60 per cent reduction of sulphur dioxide emissions below 1980 levels by 1995 – would require that twelve CEGB power stations be retrofitted with FGD systems. There is little alternative other than to retrofit, given that there are no power stations nearing the end of their lifespan which could be replaced by power stations incorporating the latest stringent pollution-control technology. The government's policy of increasing electricity generation using nuclear power (so allowing the shut-down of one or more coal-fired power stations) is unlikely to be implemented quickly enough to provide an alternative. Given the public's reaction to the Chernobyl nuclear reactor accident, the nuclear-power programme is expected to face further delay and it is likely that two more coal-fired power stations will have to be built (previously, none were planned this century) to fill the gap caused by the delay in establishment of additional nuclear power stations. The new coal fired power stations will take between seven and eight years to complete and they will have to incorporate FGD systems.

The overall capital cost of retrofitting FGD to the twelve largest CEGB coal-fired power stations is estimated to be £1432 million at 1983 prices over ten years, although if the output capacity lost through the use of the FGD system were to be replaced, the total would reach £1990 million. This figure compares with the total cost of implementing the 1956 and 1968 United Kingdom Clean Air Acts over the period 1958–84 of £757 million at 1983 prices (UK House of Commons Environment Committee, 1984). However, the financial impact of FGD would not occur at one time. An investment programme to introduce FGD at existing power stations would result in capital and operating expenditure of up to £450 million per annum (1984 prices). For the consumer the whole programme would add 6 per cent or less to electricity prices (UK House of Lords Select Committee on the European Communities, 1984).

Perhaps because retrofitting FGD at CEGB power stations is seen to be only a matter of time, given the intense EC pressure placed on the United Kingdom – or perhaps because sulphur dioxide emissions might begin to increase between now and the end of the century – in 1986 the CEGB announced its intention to retrofit FGD to three of its largest stations. This action will be phased over the period 1989–97 and will cost £600–780 million, adding 1–2 per cent onto the price of electricity. Whilst other European nations welcomed Britain's intended action, many thought that it would prove to be of limited effectiveness, especially as one of the power stations to be fitted with FGD (Drax B in Yorkshire) was only completed three months prior to the CEGB announcement, and this station would have increased CEGB sulphur dioxide emissions anyway (figure 11.4).

In contrast to the United Kingdom, West Germany, with massive and continuing public and political pressure for action to be taken concerning 'Waldsterben' (death of the forests), imposed emission control regulations in 1983 even more severe than those required by the EC Directive. All major power stations are required to reduce sulphur emissions by 85 per cent and oxides of nitrogen by 60 per cent in half the time-period specified by the EC Directive. The cost will add about 10 per cent to consumer electricity prices. Not only will the West German programme produce national benefits in the form of a reduction in acid rain damage, but it has injected 'technology stimulation' into its industry, thereby producing potential benefits for the economy through increased international trade.

Figure 11.4 The CEGB's coal-fired power station, Drax B, near Selby in Yorkshire, was completed in 1986. At the time of construction, the 260 metre-high stack was the tallest in Europe. Following its completion the CEGB announced its intention to retrofit a flue gas desulphurization system
Source: UK Central Electricity Generating Board

West Germany's strict pollution-control policies have induced innovations in production processes and West Germany is now the European leader in the technical implementation of abatement techniques such as Fluidized Bed Combustion systems and Flue Gas Desulphurization techniques. In contrast, in order to meet the stringent EC pollution-control requirements, the United Kingdom is likely to have to adopt an expensive 'crash programme' involving the import of pollution-control systems and expert personnel from countries such as West Germany that have already embarked on a programme of abatement (UK Royal Commission on Environmental Pollution, 1984).

The large combustion plant Directive places more emphasis on sulphur dioxide than oxides of nitrogen, but the contribution of oxides of nitrogen to acid rain is growing rapidly in Europe, as is photochemical oxidant (ozone) production which requires the presence of oxides of nitrogen (figure 11.5). Emissions of oxides of nitrogen in West Germany, for example, increased by 85 per cent during the past decade to reach over 3 million tonnes (table 11.2), of which motor vehicles contribute about 40 per cent (Wetstone and Rosencranz, 1983). To reduce such emissions, an EC Directive requires emissions of oxides of nitrogen, carbon monoxide and hydrocarbons from new vehicles to be markedly reduced by 1995. In order to comply, larger cars will have to fit catalytic converters and smaller cars will have to use lean-burn engines. Other actions to reduce the acid rain problem which the Community may take in the near future include a limit on the sulphur content of heavy fuel-oil and coal (perhaps to 1 per cent), a tightening of the limit on the sulphur content of gas-oil, the adoption of air-quality guide values for sulphur dioxide and oxides of nitrogen as binding limit values, the adoption of emission limits for each country in the form of national emission 'bubbles' within which trading and banking of emissions may be allowed (the Netherlands already has a self-imposed national sulphur dioxide limit), and the introduction of charges for sulphur dioxide from large emission sources.

Although the EC, with its twelve Member States, is close to taking important action on the acid rain problem, it must be recognized that the EC alone will not solve the acid rain problem in Europe: similar action on a larger international scale is needed. Whereas non-member nations such as Austria, Switzerland and the Scandinavian countries will match or even better the EC action, the effective tackling of the European acid rain problem also requires the co-operation of Eastern European nations. While large quantities of pollution continue to spill over from Eastern Europe into Western Europe, the EC action will be of limited success. The next stage – the most difficult of all – is to persuade Poland, Czechoslovakia and East Germany to reduce their emissions substantially. International organizations such as UNEP and the ECE have a major role to play in achieving such an aim.

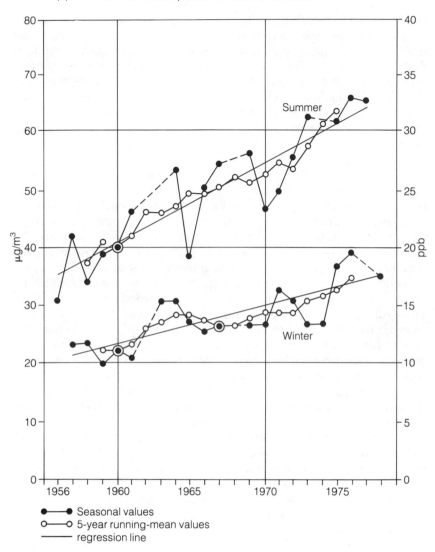

Figure 11.5 Seasonal ozone concentrations at Arkona, East Germany, 1956–78
Source: after Warmbt, 1979

11.3.5 The North American response

Co-operative action to control transfrontier pollution in North America would seem to be less problematic than in the more complex European setting, especially in the light of the long history of cordial relations between the United States and Canada. Both countries agree that they suffer pollution damage from emissions originating in the other country. About two-thirds of the sulphur deposition in eastern Canada originates from emissions in the United States. Canadian sources are, on the other hand, responsible for about one-third of the deposition in the northeast United States. The United States also imports considerable air pollution from its southern neighbour, Mexico. The United States emitted about 24 million tonnes of sulphur dioxide and 19 million tonnes of oxides of nitrogen in 1980, as compared with Canada's emissions of 5 million tonnes of sulphur dioxide and about 2 million tonnes of oxides of nitrogen. About 80 per cent of the sulphur dioxide and 65 per cent of the oxides of nitrogen in the United States originates from within the thirty one states bordering or east of the Mississippi River (Miller, 1984). Power stations account for three-quarters of the sulphur dioxide emissions in this area.

More than half of the pollution thought to be responsible for acidifying southeast Canada's lakes originates in the midwestern industrial region of the United States (Ohio, Illinois, Indiana, Kentucky, Pennsylvania and West Virginia). This region's pre-1970 power stations are generally subject to less stringent emissions standards than other US power stations which follow the strict New Source Performance Standards (NSPS) regulation of the Clean Air Act. This requires new coal-fired power stations to reduce uncontrolled sulphur dioxide emissions by between 70 and 90 per cent, depending upon the sulphur content of the coal used. This means that in effect, new power stations have to install FGD systems. Such controlled coal-fired power stations emit on average seven times less than older power stations regulated only under the sulphur dioxide air-quality standards (Wetstone and Rosencranz, 1983). Obviously, as existing sources are retired and replaced with facilities more stringently regulated by the NSPS of the Clean Air Act, emissions will be reduced. Unfortunately, it may take twenty or thirty years before newer, cleaner industrial plants are introduced, and so action is needed in the meantime to reduce acid rain. In particular, much attention has been directed towards imposing tighter emission controls on power stations, especially those in the Midwest.

During the late 1970s Canada began to express alarm at the increasing acidification of many thousands of lakes in eastern Canada and the expected harsh impact that this would have on tourism. Further concern was expressed regarding possible damage to its forests, given that one in ten of the population is employed in forestry. Consequently, in 1980 the United States and Canada entered into a Memorandum of Intent to 'develop a bilateral agreement that will reflect and further the development of effective domestic control programs and other measures to combat transboundary air pollution'. In 1982 Canada offered to reduce its sulphur dioxide emissions by 50 per cent if a comparable reduction programme was mandated in the United States. Canada waited for the US to commit itself, but the 50 per cent emission reduction programme was formally rejected (Giles, 1984). Not only did the United States walk away from the negotiations, it worsened the acid rain situation by relaxing many emission-control requirements for midwestern power plants. Not surprisingly, this caused some difficulties in traditionally cordial Canadian-US relations. In 1984 Canada decided to reduce unilaterally its 1980 sulphur dioxide emissions by one half by the year 1994.

The reasons for the unwillingness of the United States to commit itself to an emission reduction programme of 50 per cent (or even to join the Thirty-Per-Cent Club by signing the LRTBAP Convention protocol) were similar to those put forward by the United Kingdom. The United States points out that through its Clean Air Act, it has reduced its sulphur emissions by 36 per cent over the last 15 years, and at considerable cost (Rosencranz, 1986). It was not willing to require industry to pay out further vast sums of money on emission controls while doubt still existed as to whether this action would lead to the recovery of damaged aquatic ecosystems and forests. The need for greater certainty concerning the cause and effects of acid rain led to large sums being spent on further research including $250 million by the Reagan Administration (Rosencranz, 1986). However, co-operation on the acid rain problem between Canada and the United States did not cease completely. Special envoys on acid rain were designated by the respective governments, but it is clear that the Reagan Administration expected these talks to result only in an agreement to a modest $1 billion a year control programme which would reduce sulphur dioxide emissions by 1 or 2 million tonnes (Lewis and Davis, 1986). This is far short of the 50 per cent reduction in national emissions which Canada and scientific groups such as the US National Research Council (1983a) have urged.

Although the Reagan Administration placed an emphasis on requiring further research before adopting any major emission-reduction programme, several politicians have attempted to tackle the acid rain problem by strengthening aspects of the Clean Air Act. Unfortunately, they have failed to get very far, even though the Clean Air Act (section 115) includes a means by which the EPA may require that states revise their air-quality plans to eliminate emissions that 'cause or contribute to air pollution which may reasonably be anticipated to endanger public health or welfare in a foreign country' (Wetstone and Rosencranz, 1983). The total amount of emission reduction typically debated ranges from about 8 to 12 million tonnes of sulphur dioxide. Specifically, states may be required to develop strategies which produce a 5 million tonne reduction of sulphur dioxide emissions by the early 1990s and a further identical reduction by the late 1990s. By selecting, say, fifty of the largest sulphur dioxide emission sources and requiring them to retrofit FGD systems, this could achieve a reduction of 7 million tonnes (Abeles, 1984). The cost would amount to between $3.8 and 4.9 billion a year, compared with acid rain costs of at least $5 billion annually.

Although the Reagan Administration resisted the adoption of a major emission reduction programme, it seems that adoption of such a programme is only a matter of time. Nevertheless, every year that passes without a decrease in emissions of the precursors of acid rain means increased severity and extent of acid rain effects in North America. Damage to aquatic ecosystems over a wider area (e.g. southern and western United States) is already being documented and the rapidity with which forest die-back occurred in Europe should act as a warning to the United States. Even so, it seems likely that effective action to reduce acid rain will be undertaken by Europe – specifically, the European Community – before it is taken in North America, even though the situation in Europe is far more complex.

11.4   Stratospheric Ozone Depletion

Whereas the damage caused by acid rain is well documented, the effects of stratospheric ozone depletion – namely, harm to plants, animals and people by the resulting increased UV-B radiation, as well as a change in the Earth's surface temperature (refer to chapter 6) – have not yet been realized. Indeed, conclusive evidence that stratospheric ozone is being depleted is not yet available. It is therefore somewhat unexpected to find that precautionary control measures have already been taken against this potential pollution problem by many nations. This response, and the short time that has elapsed since the first indication that an air-pollution threat existed (in 1974) before effective action was taken (in 1978 by the United States), contrasts with the rather reluctant and slow response to many other pollution problems requiring international collaboration.

It is the chlorine-containing substances, especially the chlorofluoro-carbons F-11 and F-12 (accounting for almost 90 per cent of CFC production), which have been the subject for emission reduction by various countries. These two CFCs are used as propellants in aerosol sprays, as foam-blowing agents, and as working fluids in refrigeration systems. The first country to translate its concern about the CFCs' depletion of the levels of stratospheric ozone into action was the United States which in 1978 initiated a ban on non-essential use of CFCs in aerosol propellants (for example, deodorants, polishes, hair lacquer and paint) – the ban becoming fully effective in 1979 (Brennan, 1979; Stoel, 1983). Given that the United States was the world's largest producer of CFCs (40 per cent of world sales of CFCs in 1976), then its willingness to act persuaded other nations with a strong global environmental conscience to follow this lead. A number of other countries, including Canada, Denmark, Finland, Norway and Sweden, joined the United States (collectively referred to as the Toronto Group after an informal negotiating meeting held in Canada) in imposing a ban or a restriction on non-essential aerosol uses of CFCs.

The United States also provided the initiative which led to the United Nations Environment Programme (UNEP) forming a Co-ordinating Committee on the Ozone Layer (CCOL) in 1977. UNEP is one of the largest international organizations concerned with environmental issues. It was set up following the United Nations Conference on the Human Environment held in 1972. The United Nations General Assembly elects fifty eight member-governments to serve as the Governing Council. The major goals of UNEP are to increase knowledge of the environment and to protect and improve its quality – that is, to develop an 'Earthwatch' function. Specifically, UNEP was charged to keep the world environmental situation under review, in order to ensure that emerging environmental problems of wide international significance received appropriate and adequate consideration (Holdgate et al., 1982, 1982b). As part of these developments, UNEP has established the Global Environment Monitoring System (GEMS), in co-operation with the World Health Organization, which has been designed to assess air-pollution conditions on a global scale, to observe trends, and to begin to examine the relationship between pollution and human health (Bennett et al., 1985; GEMS, 1984). EMEP forms part of GEMS. Information exchange on national environmental information is achieved through INFOTERRA, which involves 112 countries and is co-ordinated by UNEP at its headquarters in Nairobi (Clarke and Timberlake, 1982). A wide range of environmental information can be suplied by UNEP, such as details from the International Register of Potentially Toxic Chemicals, which lists the health and environmental effects of chemicals, as well as the regulatory measures taken by individual countries to control these chemicals.

In contrast with the partial ban on CFCs initiated by the United States, the European Community, responsible for 34 per cent of the world sales of CFCs in 1976, issued a Directive in 1980 which placed a ceiling on the total Community production of CFCs and required a 30 per cent reduction in the use of CFCs in aerosol cans by the end of 1981, as compared with 1976 levels. Member States were also required to develop the best practicable technologies in order to limit emissions in the refrigeration, solvent and foam-plastics sectors. CFCs in aerosols accounted for about 65 per cent of total CFC production in the EC. At one stage the European Parliament had debated a 50 per cent reduction by 1981 and a total ban (except for essential medical purposes) by 1983 (Haigh, 1984). Nevertheless, within this period a reduction of 35 per cent was achieved, although it is uncertain to what extent the achievement of the target resulted from the Directive on the one hand, or from economic factors coupled with scientific and public concern about stratospheric ozone depletion on the other.

By 1983, as a result of action taken by European and North American countries, there was a 21 per cent reduction in the world production of

CFCs F-11 and F-12 from the peak level of 1974 (Sand, 1985). However, this statistic does conceal that production of F-11 and F-12 increased by 7 per cent between 1982 and 1983. Also, non-propellant uses of CFCs are growing quickly and it is clear that both Eastern Europe and India have increased production of CFCs substantially during the 1980s (UK Royal Commission on Environmental Pollution, 1984). This highlights the fact that the action taken to date may have been only of short-term importance because the global production of CFCs is beginning to increase again.

The need for a truly worldwide international collaboration resulted in UNEP-CCOL arranging a conference on the ozone issue for member countries in Vienna in March 1985. The outcome was that twenty participating countries and the European Community (on behalf of its Member States) signed the 'Vienna Convention for the Protection of the Ozone Layer' (table 11.4). The stated purpose of this framework

Table 11.4 Signatories to the Vienna Convention for the Protection of the Ozone Layer

State	Date of signatures	
Argentina	22 March	1985
Belgium	22 March	1985
Byelorussian Soviet Socialist Republic	22 March	1985
Canada	22 March	1985
Chile	22 March	1985
Denmark	22 March	1985
Egypt	22 March	1985
Finland	22 March	1985
France	22 March	1985
Germany (Federal Republic)	22 March	1985
Greece	22 March	1985
Italy	22 March	1985
Netherlands	22 March	1985
Norway	22 March	1985
Peru	22 March	1985
Sweden	22 March	1985
Switzerland	22 March	1985
Ukrainian Soviet Socialist Republic	22 March	1985
Union of Soviet Socialist Republics	22 March	1985
United States of America	22 March	1985
European Economic Community	22 March	1985
Mexico	1 April	1985
Luxembourg	17 April	1985
United Kingdom	20 May	1985
Austria	16 September	1985
Morocco	7 February	1986
Burkina Faso	12 December	1986

Source: Information supplied by the United Nations Environment Programme Headquarters, Nairobi, Kenya

Convention (or umbrella treaty) is to promote exchanges of information, research, and data on monitoring with the aim of protecting human health and the environment against activities that have an adverse effect on the ozone layer (defined as the layer of atmospheric ozone above the planetary boundary-layer). It spells out the general responsibility of signatories for preventing environmental harm caused by human interference with the ozone layer. It creates new, permanent, international institutions to implement these rules and to elaborate more specific rules in the form of protocols, technical annexes, and financial and procedural provisions. The Convention represents a major breakthrough in involving non-Western nations in tackling the ozone depletion threat. However, it is only the first step towards a world response to a global issue: it remains to be seen what effective action will be taken to reverse the rising global production of CFCs, for no specific emission-control measures were adopted.

Given that European and North American countries had already taken some regulatory measures, it was disappointing that the Vienna Conference failed to produce a Convention protocol specifying further emission control of CFCs and involving more countries. The main reason why no agreement was reached was that the Toronto Group and the European Community disagreed on control options. The Toronto Group wanted all countries to cut back on non-essential uses of CFCs as they had already done so. In contrast, the European Community advocated the specification of a production cap on F-11 and F-12, and a 30 per cent reduction in non-essential uses of CFC aerosols, as it had already done so. Both opposed each other's approach since this would mean one group not having to make any further reductions, while the other group would be subject to new restrictions which might affect their export industries (Sand, 1985).

One solution to the stalemate is to set a global cap on emissions, with national quotas for CFC production being fixed as proportions of population size and Gross National Product. This would allow each country to decide to what uses CFCs were put. The initial global ceiling might not be severely limiting but it could be tightened in the near future if international concern for ozone depletion by CFCs remains high or even increases. If no firm action is taken world production of CFCs may continue to increase. Considering that recent predictions point to a depletion level of 3–5 per cent in ozone being reached after 50 years or more, as opposed to the 15–20 per cent predicted by research in 1980, then it may be that some countries are reluctant to endorse strong regulatory measures. Generally, it appears that further progress towards effective emission control at a global scale will be slow.

Even while further international emission control of CFCs F-11 and F-12 is being debated, other chlorine-containing gases are becoming of increasing potential significance for ozone depletion (US National Research Council, 1984). It may be that any regulatory action will have to consider limiting emissions of such gases as hydrogen chloride released directly into the stratosphere by the US Space Shuttle, methylchloroform (CH_3CCL_3) used as an industrial non-toxic metal-cleaning solvent, carbon tetrachloride (CCL_4) used as a dry-cleaner, as well as CFC F-22 ($CHCLF_2$) and methyl chloride (CH_3CL). The issue of stratospheric-ozone depletion by pollutants is thus becoming far more complex than was once believed to be the case.

11.5   Carbon Dioxide Build-up

Another global pollution issue in which UNEP has played an important role concerns the increase in carbon dioxide concentration in the atmosphere arising from increased fossil-fuel burning and land-use changes. This may result in changes in the global climate and atmospheric circulation, thereby producing far-reaching effects for humanity, especially in the next century. UNEP has arranged a number of international meetings on this issue, including the conference (convened jointly with the World Meteorological Organization and the International Congress of Scientific Unions) held in Austria in 1985. This conference involved scientists from twenty-nine developed and developing nations in assessing the role of increased carbon dioxide and other radiatively active constituents of the atmosphere (collectively known as greenhouse gases) on climate changes and associated impacts. It was agreed that if present trends continue, then the combined concentrations of greenhouse gases will be radiatively equivalent to a doubling of the amount of carbon dioxide from pre-industrial levels, possibly as early as the 2030s, and this is likely to lead to an increase in the Earth's surface temperature of between 1.5° and 4.5°C (refer to section 6.2). In addition, it may increase ozone by between 3 and 6 per cent and so have profound implications for the stratospheric-ozone depletion issue.

The implications of a global warming, together with associated changes in atmospheric circulation systems, are far-reaching, but governments have not yet altered their energy policies to reduce carbon dioxide emissions. Whereas several countries were willing to take precautionary action on the perceived ozone-depletion threat, since it affected only a few selected (several non-essential) industries, governments are far more reluctant to consider changing their energy policies. Given this situation, the role of organizations like UNEP is to increase worldwide awareness of the potential seriousness of the pollution problem, and to urge governments to keep all energy options open and to keep energy planning as flexible as possible so as to cope with uncertainty. The possible changes in climate that may be a consequence of an increasing concentration of carbon dioxide need to be considered when governments plan long-term projects such as irrigation and hydro-electric power, drought relief, agricultural land-use, structural designs and coastal engineering works, and energy development. It is important that such projects retain a degree of flexibility so that they can be adapted to significant changes in the climate.

When governments have recognized the full consequences of this pollution problem, hopefully before the effects are manifest, UNEP may attempt to co-ordinate the signing of a global framework Convention, like the Vienna Ozone Convention, to mark an international commitment to the initiation of a long-term programme of action. The subsequent action needed would be to reduce emissions of carbon dioxide by reducing energy consumption (such as by the more efficient use of present energy supplies and taxation on the use of fossil fuel) and reducing our reliance on fossil fuels (by increasing the use of renewable sources of energy, or even nuclear power). By far the greatest contribution to atmospheric carbon dioxide comes from the burning of coal, and this is likely to remain so for the forseeable future. Over 60 per cent of the world reserves of coal are in Western Europe, the United States and the Soviet Union. In principle, therefore, the governments of a relatively small number of countries hold the key to international action to reduce carbon dioxide emissions. It is significant that it is the grain-growing regions of these very countries whose agricultural productivity is likely to be most affected by a change in the climate (UK Royal Commission on Environmental Pollution, 1984).

As carbon dioxide is also being added to the atmosphere by land-use changes, an international commitment to stopping the widespread destruction of forests, especially in low latitudes, would at least lessen the carbon dioxide build-up. Extensive planting of trees, increasing the storage of carbon in the biosphere, is another form of action that can slow down the increase in carbon dioxide concentration in the atmosphere. However, to account for 2.5 Gt of carbon a year, an area of 3 million square kilometres of new forest would be needed, which is equivalent to the size of India, or 6–10 per cent of the present world forest area. It seems highly unlikely that such an extensive planting scheme could be initiated, and it is probable that there would be considerable pressures in favour of increased food production when decisions on land-use came to be made (Liss and Crane, 1983). Another part of the biosphere which some researchers have suggested could increase its store of carbon is the marine organism. An increased supply of phosphates and nitrates to surface waters could fertilize the growth of marine organisms which would incorporate carbon dioxide and eventually sink to the ocean floor.

By analogy with the removal of sulphur dioxide from power-station emissions, it has been suggested that carbon dioxide could be removed from the flue gases of major emission sources. The carbon dioxide removed would have to be disposed of in such a way that it would be a long time before the carbon dioxide returned to the atmosphere. Marchetti (1977) proposed the injection of the collected carbon dioxide into seawater at locations such as the Straits of Gibraltar, where water sinks to great depths in the ocean and does not reach the surface again for a thousand years or more. Another suggestion is that of using monoethanolamine to extract the carbon dioxide from the flue gases. The monoethanolamine is then regenerated and the carbon dioxide which it has absorbed is compressed and transported to the coast for dumping into the deep oceans (Baes et al., 1980). However, at the present time the removal of carbon dioxide from flue gases and its disposal remains technically uncertain and excessively costly. In the meantime, UNEP stresses the need to increase international co-operation in monitoring the situation (climatic change and carbon dioxide levels), to increase understanding of the implications of the predicted climatic change for each nation and for world society, and to increase awareness of what international action may be needed. UNEP is also urging nations to make more efficient use of present energy supplies, as this would not only reduce carbon dioxide emissions but would also reduce the consumption of fossil fuels.

11.6 International Progress to Date

The discussion of acid rain, stratospheric ozone depletion and carbon dioxide build-up highlights the fact that progress towards international co-operative action on any pollution issue occurs slowly and hesitantly, if at all. It usually takes a lengthy period of international discussion before nations agree that a pollution problem is serious enough to warrant the taking of action, but even then the introduction of control measures may not automatically follow. International organizations such as OECD, the ECE and UNEP play an important role in stimulating national awareness and understanding of pollution problems which require an international response, and in formulating appropriate responses to these problems. What such organizations have been less successful in doing is converting this heightened concern by national governments into effective co-operative regulatory action. When it comes to the need for emission-reduction programmes, most nations display a reluctance to commit themselves to action which may be economically costly and socially disruptive.

The degree of reluctance or willingness of a country to take action on a pollution problem depends upon many factors, such as the strength of support for environmental issues among its public, the perceived seriousness of the pollution problem for itself, the state of its economy, and even the timing of a proposal for international action. Whereas countries with a strong environmental consciousness, such as the Scandinavian nations and to some extent Canada, may be ready to act swiftly on a variety of pollution problems, there are other countries which appear to express reservations concerning the need to act at all.

Surprisingly, in view of its historical lead in introducing national pollution-control legislation, the United Kingdom is listed among the latter group of countries. Frequently, the United Kingdom urges the need for greater certainty concerning the causes and nature of a pollution problem and the benefits which would result from pollution-control measures before it is willing to commit itself to regulatory action. Generally, the level of uncertainty which a government will accept before contemplating action depends upon the perceived seriousness of the problem for that country. Thus, the sudden discovery by West Germany that it was suffering seriously from acid rain (and photochemical oxidant episodes) created massive internal public and political pressure on the government to change its policy towards acid rain control measures. Whereas in 1979 it opposed regulatory action on acid rain, in 1982 it became a leading force in urging that effective international action be taken. It is even possible that a similar transformation in response to the acid rain problem may take place in the United Kingdom and the United States in the near future. A change in attitude may also follow a major pollution accident, such as occurred at Seveso, Bhopal and Chernobyl, because governments are confronted by strong internal public and political pressure to be seen to be taking action. Even when a country is willing to act on one pollution problem, and becomes credited with being the motivating force for initiating international action on that issue, it may be reluctant to take action where another pollution problem is concerned. Thus, for example, while the United States initiated the international response to the stratospheric-ozone depletion problem, it remains reluctant as of yet to take effective action on acid rain.

Given that pollution-control measures are economically costly to introduce, many nations are reluctant to act because they believe that such action would slow down their rates of industrial and economic growth: for example, industrial nations facing economic hardship such as Poland and many developing countries with limited resources are reluctant to divert funds towards pollution control. It is with regard to this attitude that international organisations need to stress to nations two important considerations. Firstly, pollution control is an integral part of long-term economic growth since it aims to safeguard a national potential for future development by preventing damage and wastage of scarce natural resources. International organizations need to encourage nations to recognize the economic benefits of reducing and preventing environmental damage. Reassuringly, an increasing number of developing nations now realize that their ultimate goals cannot be reached without sound environmental management (Clarke and Timberlake, 1982).

With regard to the second consideration, it is now widely accepted that economic growth is considered essential to progress being maintained in environmental management. Economic recession and poverty often lead to air-quality deterioration as funding for expenditure on pollution control competes with other needs. Economic progress in the developing world is likely to be the only means whereby developing countries gain the resources for pollution control. Whether wealthier nations would be willing to contribute to an international fund to help those countries which lack adequate funds to introduce effective pollution-control measures remains to be seen. Both considerations point to a need to anticipate pollution problems more accurately and so provide time to integrate pollution-control policy into economic, social, energy, industry, agriculture, transportation, and land-use policies. Considerable time is needed to introduce industries and technologies which produce no, or very little, waste. Nevertheless, this is the ultimate pollution-control policy since a reduction in wastes will reduce the pollution problems that have to be faced in the future.

Many countries are reluctant to act without scientific proof of the existence of a pollution problem being available – as, for example, in the case of stratospheric-ozone depletion and carbon dioxide build-up. It is a difficult decision for nations to take, as to whether to act now on an assumed pollution threat or to wait for further scientific proof. The danger is that while more and better research is desirable for any problem, any delay in taking action on some pollution problems may only increase the severity of the pollution effects when they are realized. Nations are then faced with taking remedial action at considerable economic and social cost. In such a situation, anticipative action is far more effective than remedial action, but this does mean that some degree of uncertainty as to the reality of a potential pollution threat has to be accepted by nations. In this connection, progress on the issue of stratospheric-ozone depletion by CFCs is enlightening. Exceptionally, precautionary regulatory action quickly followed the highlighting of a global pollution threat. Although this example may be used to highlight the desirable speed of response concerning other global pollution problems, it must be pointed out that the action taken was not sufficiently co-ordinated internationally to continue the initial progress made. Action taken by the Toronto Group of nations and the European Community was partially offset by those countries which did not take action. Moreover, some of those nations now point out that research during the past decade has successively reduced the estimates of the seriousness of the stratospheric-ozone depletion threat. Regrettably, should subsequent research reveal that there never was a problem, then it could influence how nations react to other global pollution threats, such as the adverse effects predicted to occur as a result of the increase in carbon dioxide in the atmosphere.

International organizations not only promote international collaboration on transfrontier and global pollution problems, but they also attempt to make governments more aware of their responsibilities to their own environments. Declarations such as that agreed at the United Nations Conference on the Human Environment in 1972 do assist in achieving this. As national policies are developed or modified to reflect a nation's obligation to its own environment, it will in turn help to reduce transfrontier and global pollution problems. Thus, for example, policies which reduce the emissions of pollutants, whether through the application of air-quality standards or emission standards, reduce the volume of wastes which may pass beyond a country's frontiers. Many nations apply policies (excluding the tall-stacks policy) which have successfully reduced the concentrations of suspended particulate matter and gaseous pollutants in urban areas, and this has meant a reduction in the potential pollution which may be transported across frontiers and around the world.

Considering that transfrontier and global pollution problems have only received serious international attention since the 1970s, then whilst one may express disappointment over the rate of progress, at least some progress on international collaboration has been made. International Conventions such as those agreed on stratospheric ozone and acid rain are important first steps towards the goal of effective regulatory action on these problems. What is now needed is a greater international commitment to transforming awareness and concern for pollution problems into effective and lasting action. As Holdgate et al. (1982a) point out in relation to international progress on pollution control, 'the ratio of words to action is weighted too heavily towards the former'.

Appendix 1
Prefixes for SI units

Prefix	Symbol	Factor
tera	T	10^{12}
giga	G	10^{9}
mega	M	10^{6}
kilo	k	10^{3}
milli	m	10^{-3}
micro	μ	10^{-6}
nano	n	10^{-9}

Appendix 2
The Stratospheric Ozone Hole

1 The Discovery of the Ozone Hole

In 1985 measurements at Halley Bay on the Antarctic coast were published revealing that extremely large stratospheric ozone depletions had occurred there each spring since the late 1970s (Farman et al., 1985). Subsequent measurements using the US Nimbus-7 satellite, aircraft and balloon-borne instruments confirmed ozone losses up to 50 per cent in some (southern) spring months in the lower Antarctic stratosphere. Whereas Antarctic October total ozone levels were between 280 and 330 Dobson Units during the 1957–76 period they dropped to 225–50 by the early 1980s and to less than 200 in the following years. Each Dobson Unit is a hundredth of a millimetre and refers to the thickness of the stratospheric ozone layer that would result if it were brought to sea level pressure and temperature.

The reduction in ozone concentration at altitudes between 12 and 25 km begins with the first sunlight of the Antarctic spring in late August and is greatest through September and October. In October 1988 a 25 per cent drop in ozone levels was measured by the Nimbus-7 satellite while in October 1987 the depletion was about 50 per cent. Satellite observations reveal the area of major ozone loss to be about the size of the United States. Immediately outside the ozone-depleted zone is often a crescent-shaped region of ozone-rich air which accentuates the image of an ozone 'hole' as depicted on maps of Antarctic ozone concentration. By November/ December the springtime circumpolar atmospheric vortex which has isolated stratospheric Antarctic air from the rest of the southern hemisphere breaks down. Air from outside the hole, which has not been so depleted of ozone, then enters the region and the mixing of the airmasses dissipates the hole (Rowland, 1988).

The shape and position of the hole varies from year to year. The southern tip of South America is the only landmass outside Antarctica above which the large ozone depletions are evident. For example, during October 1987 a 13 per cent stratospheric ozone reduction was measured there. However, given that the hole is primarily located over Antarctica, the direct effects of allowing increased UV-B radiation to reach the ground would appear to be minor at the moment, for humans at least. The same may not apply to the marine ecosystems of the oceans surrounding Antarctica.

The discovery of the Antarctic hole, at a time when global trends in ozone levels were uncertain, came as a complete surprise. It had not been predicted by any of the models simulating potential pollution effects on the stratosphere. To explain why the ozone hole exists, research now suggests the Antarctic region offers an extreme situation in which chlorine molecules can scavenge ozone at a very fast rate. The key to this appears to be the presence of high-altitude ice clouds in the lower stratosphere where temperatures dip below $-80°C$. It is suggested nitrogen compounds in the air become bound up with the cloud ice particles, temporarily removing any free nitrate which would otherwise react and hold the ozone-depleting chlorine as chlorine nitrate. It is this formation of chlorine nitrate in the global stratosphere which is the reason why researchers believe the ozone-depleting potential of chlorine is not as great as was originally feared on the worldwide scale. In contrast, the Antarctic stratosphere provides conditions which prevent the usual reactions between chlorine and nitrogen compounds. In addition, chemical reactions on the surfaces of the ice cloud particles may speed up the rate of ozone depletion by some forms of chlorine (Stolarski, 1988). If the global pollutant loading of chlorine continues to build up, ozone destruction will proceed even more rapidly, and the Antarctic hole is likely to appear deeper and earlier each year.

The atmospheric circulation and the warmer conditions in the Arctic stratosphere appear slightly less favourable for producing an ozone hole. Even so, reductions in ozone levels have been recorded there recently in the spring, especially during April (Heath, 1988). On the global scale, although natural fluctuations due to solar sunspot cycles complicate any assessment of ozone trends, a reanalysis of data for the 1979–86 period suggests a global ozone loss of about 2.5 per cent or 0.35 per cent per year.

2   Progress towards Effective International Action
    on regulating CFCs

The signing of the Vienna Convention for the Protection of the Ozone Layer in 1985 was the first international step towards action to protect the ozone layer, but it did not provide for any specific emission-control measures. Subsequent discussions concerning what international actions should be taken appeared to be making little progress. The situation changed with confirmation of the existence of a springtime ozone hole. A sense of urgency that some action had to be taken, even if it was a compromise between opposing views, began to emerge. The result was that 27 nations agreed the Montreal Protocol on Substances that Deplete the Ozone Layer in late 1987. The protocol took effect in 1989 and requires specific international actions to be taken concerning five CFCs and three halons:

1   a freeze on the production and consumption of CFCs at 1986 rates from 1990
2   a freeze on the production and consumption of halons at 1986 rates from 1992
3   a 20 per cent reduction in CFC emissions below the 1986 level from 1994
4   a 50 per cent reduction in CFC emissions below the 1986 level from 1999
5   agreement for less-developed countries to develop capabilities for CFC production for an extra decade, with limitations affecting them delayed also by a decade beyond the time-scale for the developed countries.

Currently, of the five CFCs affected, about 800,000 tonnes of F-11 and F-12 are produced annually worldwide, in about equal quantities, compared with about 170,000 tonnes of F-113, and 30 tonnes of F-114 and F-115. The protocol also covers three bromine-containing compounds, halon-1211, halon-1301 and halon-2402, which are used in fire extinguishers. Bromine catalyses the breakdown of ozone even faster than chlorine.

Despite the stated intentions of the protocol, a 50 per cent global reduction in CFC emissions by 1999 is unlikely to be achieved. Emission increases by developing countries, which at present have negligible CFC production, will counter the reductions made by the developed countries. At best, this will convert the intended 50 per cent cutback to about 35 per cent. At worst, growth in CFC production and consumption in developing countries may actually lead to a global *increase* in emissions over the next decade rather than a decrease (Rowland, 1988).

Given the depletion of stratospheric ozone that is taking place over Antarctica, and to a lesser extent globally, many researchers are suggesting the need for an *immediate* emission reduction of 85 per cent or even 95 per cent. The UNEP protocol does not prevent countries from taking stronger action than specified and some countries have committed themselves to greater cuts. Sweden and West Germany aim for near-total elimination of CFCs by the end of the century while the United States and Britain endorse a cut of 85 per cent at least. In contrast, some developed nations such as Japan, with its vast electronics industry in which F-113 is extensively used for cleaning electronic components, are reluctant to agree to such drastic reductions. Yet it appears that it is the industrial nations who will have to commit themselves to as much as 95 per cent CFC reductions if a strengthening of the UNEP protocol is to be achieved when it is reviewed in 1990 and 1994. Low CFC per capita consuming countries such as China, India, South Korea and Brazil are only likely to limit their CFC development programmes if they see the industrialized nations making such a substantial commitment to environmental protection.

Progress towards elimination of the use of ozone-depleting CFCs will ultimately depend upon the availability of substitutes to replace these compounds. The protocol has stimulated the chemical industry to develop alternatives which are 'ozone-benign' or 'ozone-friendly'. CFCs are likely to continue to be used in a wide variety of products but they will be the less ozone-damaging CFCs which break down before they reach the stratosphere. For example, refrigerators may use F-22 (which is largely destroyed in the lower atmosphere because it contains hydrogen) or F-134a (which contains no chlorine). However, CFCs are greenhouse gases and are contributing to a global atmospheric warming. Indeed, each molecule of CFC is far more effective at absorbing infrared radiation emitted by the Earth's surface than carbon dioxide. Given this situation, the ideal action is to develop technologies and products which do not rely on the use of CFCs at all. Increased awareness by the public of the environmental consequences of CFCs and the expression of this in consumer choice of products could be an important factor in achieving further progress in reducing the use of ozone-depleting compounds and greenhouse gases. Public concern together with campaigns by environmental pressure groups, 'green' political parties and the media may provide the political pressure needed to encourage governments to commit themselves to a stronger UNEP protocol.

References

Abeles, C. C. 1984: Current legislative directions. In T. C. Elliott and R. G. Schwieger (ed.), *The Acid Rain Sourcebook* New York: McGraw-Hill, 27–36.

Adams, J. A. S., Mantovani, M. S. M. and Lundell, L. L. 1977: Wood versus fossil fuel as a source of excess carbon dioxide in the atmosphere: a preliminary report. *Science*, 196, 54–6.

Adams, R. M., Hamilton, S. A. and McCarl, B. A. 1985: An assessment of the economic effects of ozone on US agriculture. *J. Air Pollut. Control Ass.*, 35, 938–43.

Adams, W. C. and Schelegle, E. S. 1983: Ozone and high ventilation effects on pulmonary function and endurance performance. *J. Applied Physiology*, 55, 805–12.

Aleksandrov, V. V. and Stenchikov, G. L. 1983: On the modelling of the climatic consequences of nuclear war. In *The Proceedings on Applied Mathematics*, Moscow: Computing Centre of the USSR Academy of Sciences.

Anderson, F. R., Kneese, A. V., Reed, P. D., Taylor, S. and Stevenson, R. B. 1977: *Environmental Improvement Through Economic Incentives*. Baltimore: Johns Hopkins University Press.

Anderson, I., Lundquist, G. R., Jensen, P, L. and Proctor, D. F. 1974: Human response to controlled levels of sulphur dioxide. *Arch. Environ. Health.* 28, 31–9.

Angell, J. K. and Korshover, J. 1980: Update of ozone variations through 1979. In J. London (ed.) *Proc. of the Quadrennial International Ozone Symposium, Boulder, Colorado, August 4–9, 1980*, 393–6.

Annest, J. L., Mahaffey, K. R., Cox, D. H. and Roberts, J. 1982: Blood lead levels for persons 6 months to 74 years of age: United States, 1976–80. *Advance Data from Vital and Health Statistics*, No. 79, US Dept. of Health, and Human Services Public No. (PHS) 82–150. Hyattsville, Maryland: Public Health Service.

Anonymous 1982: EMEP. The Co-operative Programme for Monitoring and Evaluation of Long-Range Transmission of Air Pollution in Europe. *Economic Bulletin for Europe*, 34, 29–40.

Anthony, R. 1982: Polls, pollution and politics: trends in public opinion on the environment. *Environment*, 24, 14–20, 33–4.

Apling, A. J., Keddie, A. W. C., Weatherley, M. -L. P. M. and Williams, M. L. 1977: *The High Pollution Episode in London, December 1975*. Stevenage: Warren Spring Laboratory, Report LR 263 (AP).

ApSimon, H. M., Goddard, A. J. H. and Wrigley J. 1985: Long-range atmospheric dispersion of radioisotopes – I. The MESOS Model. *Atmospheric Environ.*, 19, 99–111.

Ashby, E. 1975: Clean air over London. *Clean Air*, 5, 25–30.

Ashby, E. and Anderson, M. 1982: *The Politics of Clean Air*. Oxford: Oxford University Press.

Atkinson, B. W. 1968: A preliminary examination of the possible effect of London's urban area on the distribution of thunder rainfall 1951–60. *Trans. Inst. Brit. Geogr.*, 44, 97–118.

Atkinson, B. W. 1969: A further examination of the urban maximum of thunder rainfall in London, 1951–60. *Trans. Inst. Brit. Geogr.*, 48, 97–119.

Atkinson, B. W. 1970: The reality of the urban effect on precipitation: a case-study approach. *WMO Technical Note*, 108, Geneva: World Meteorological Organization, 344–62.

Atkinson, B. W. 1971: The effect of an urban area on the precipitation from a moving thunderstorm. *Bull. Amer. Met. Soc.*, 10, 47–55.

Atkinson, B. W. 1975: The mechanical effect of an urban area on convective precipitation. *Occasional Paper, Dept. of Geography, Queen Mary College, University of London*, 3.

Auliciems, A. and Burton, I. 1973: Trends in smoke concentrations before and after the Clean Air Act of 1956. *Atmospheric Environ*, 7, 1063–70.

Babich, H., Davis, D. L. and Stotzky, G. 1980: Acid precipitation: causes and consequences. *Environment*, 22, 6–13, 40.

Bach, W. 1972: *Atmospheric Pollution*. New York: McGraw-Hill.

Bach, W. 1978: The potential consequences of increasing carbon dioxide levels in the atmosphere. In J. Williams (ed.), *Carbon Dioxide, Climate and Society*, Oxford: Pergamon Press, 141–67.

Bach, W. 1979: Impact of increasing atmospheric carbon dioxide concentrations on global climate: potential consequences and corrective measures. *Environ. International*, 2, 215–28.

Baes, C. F., Beall, S. E. and Lee, D. W. 1980: The collection, disposal and storage of carbon dioxide. In W. Bach, J. Pankrath, and J. Williams (eds), *Interactions of Energy and Climate*, Dordrecht: D. Reidel, 495–519.

Baes Jr, C. F., Goeller, H. E., Olson, J. S. and Rotty, R. M. 1977: Carbon dioxide and climate: the uncontrolled experiment. *American Scientist*, 65, 310–20.

Ball, D. J. 1984: Environmental implications of increasing particulate emissions resulting from diesel engine penetration of the European automobile market. *Science of the Total Environ.*, 33, 15–30.

Ball, D. J. and Armorgie, C. J. 1983: Implications for London of European air quality standards for sulphur dioxide and suspended particulates. *Environ. Pollution (Series B)*, 5, 207–32.

Ball, D. J. and Bernard, R. E. 1978a: An analysis of photochemical incidents in the Greater London area with particular reference to the summer of 1976. *Atmospheric Environ.*, 12, 1391–401.

Ball, D. J. and Bernard, R. E. 1978b: Evidence of photochemical haze in the atmosphere of Greater London. *Nature*, 271, 733–4.

Ball, D. J. and Caswell, R. 1983: Smoke from diesel-engined road vehicles: an investigation into the basis of British and European emission standards. *Atmospheric Environ.*, 17, 169–81.

Ball, D. J. and Hume, R. 1977: The relative importance of vehicular and domestic emissions of dark smoke in Greater London in the mid-1970s, the significance of smoke shade measurements, and an explanation of the relationships of smoke shade to gravimetric measurements of particulate. *Atmospheric Environ.*, 11, 1065–73.

Ball, D. J. and Laxen, D. P. H. 1986: *Ambient Ozone Concentrations in Greater London 1975 to 1985.* London: London Scientific Services.

Barnes, R. A. 1979: The long range transport of air pollution: a review of European experience. *J. Air Pollut. Control Ass.*, 29, 1219–35.

Barnes, R. A. 1986: Acid rain: media myth or environmental apocalypse? *Catalyst*, 2, 55–70.

Barnes, R. A. and Lee, D. O. 1978: Visibility in London and the long distance transport of atmospheric sulphur. *Atmospheric Environ.*, 12, 791–4.

Barrie, L. A. 1986: Arctic air pollution: an overview of current knowledge. *Atmospheric Environ.*, 20, 643–63.

Barrie, L. A., Hoff, R. M. and Daggupaty, S. M. 1981: The influence of mid-latitudinal pollution sources on haze in the Canadian arctic. *Atmospheric Environ.*, 15, 1407–19.

Barrie, L. A., Whelpdale, D. M. and Munn, R. E. 1976: Effects of anthropogenic emissions on climate: a review of selected topics. *Ambio*, 5, 209–12.

Bates, D. V., Bell, G. M., Burnham, C. D., Hazucha, M., Mantha, J., Pengelly, L. D. and Silverman, F. 1972: Short-term effects of ozone on the lung. *J. Applied Physiology*, 32, 176–81.

Behr, P. 1979: EPA's 'bubble' strategy. *Environment*, 21, 2–4.

Bell, J. N. B. and Cox, R. A. 1975: Atmospheric ozone and plant damage in the United Kingdom. *Environ. Pollution*, 8, 167–70.

Bennett, B. G., Kretzschmar, J. G., Akland, G. G. and de Koning, H. W. 1985: Urban air pollution worldwide. *Environ. Sci. Technology*, 19, 298–304.

Beranek, L. L. 1971: *Noise and Vibration Control.* Englewood Cliffs, New Jersey: McGraw-Hill.

Bernstein, H. T. 1975: The mysterious disappearance of Edwardian London fog. *London Journal*, 1, 189–206.

Betts, P. R., Astley, R. and Raine, D. N. 1973: Lead intoxication in children in Birmingham. *Brit. Med. J.*, 1, 402–6.

Beuchley, R. W., Bruggen, J. V. and Truppi, L. E. 1972: Heat island = death island? *Environ. Research*, 5, 85–92.

Bhumralkar, C. M. and Alich, J. A. 1976: Meteorological effects of waste heat rejection from large electric power centres. *Power Engineering*, 80, 54-7.

Binder, R. E., Mitchell, C. A., Hosein, H. R. and Bouhuys, A. 1976: Importance of the indoor environment in air pollution exposure. *Arch. Environ. Health*, 31, 277-9.

Bjorseth, A. 1983: *Handbook of Polycyclic Aromatic Hydrocarbons*. Basel: Marcel Dekker.

Blacker, S. M., Ott, W. R. and Stanley, T. W. 1977: Measurement and the law: monitoring for compliance with the Clean Air Amendments of 1970. *Intern. J. Environ. Studies*, 11, 169-85.

Blowers, A. 1984: *Something in the Air: corporate power and the environment*. London: Harper & Row.

Bodkin, L. D. 1974: Carbon monoxide and smog. *Environment*, 16, 34-41.

Boeck, W. L. 1976: Meteorological consequences of atmospheric Krypton-85. *Science*, 193, 195-8.

Bond, A. R. 1984: Air pollution in Noril'sk: a Soviet worst case? *Soviet Geography*, 25, 665-80.

Bowman, J. S. 1975: The Ecology Movement: a viewpoint. *Intern. J. Environ. Studies*, 8, 91-7.

Bowman, J. S. & Fuchs, T. 1981: Environmental coverage in the mass media: a longitudinal study. *Intern. J. Environ. Studies*, 18, 11-22.

Brady, G. L. 1983: Emissions trading in the United States: an overview and the technical requirements. *J. Environ. Management*, 17, 63-79.

Brazell, J. H. 1970: Meteorology and the Clean Air Act. *Nature*, 226, 694-6.

Brennan, R. P. 1979: A soft approach to chlorofluorocarbon regulation. *Environment*, 21, 41-2.

Brice, K. A., Derwent, R. G., Eggleton, A. E. J. and Penkett, S. A. 1982: Measurements of CCL3F and CCL4 at Harwell over the period January 1975-June 1981 and the atmospheric lifetime of CCL3F. *Atmospheric Environ.* 16, 2543-54.

Bridgman, H. A. 1981: Seasonal differences in the shortwave radiation budget at Newcastle, NSW. *Australian Geographer*, 15, 89-98.

Briggs, D. J. and France, J. 1982: Mapping noise pollution from road traffic for regional environmental planning. *J. Environ. Management*, 14, 173-9.

Brimblecombe, P. 1977: London air pollution 1500-1900. *Atmospheric Environment*, 11, 1157-62.

Brimblecombe, P. 1978: London air pollution 1500-1900. *Atmospheric Environ.*, 12, 2522-3.

Brimblecombe, P. 1982: Long term trends in London fog. *Science of the Total Environ.*, 22, 19-29.

Brodine, V. 1971: Episode 104. *Environment*, 13, 2-28.

Broecker, W. S. 1975: Climatic change: are we on the brink of a pronounced global warming? *Science*, 189, 460-3.

Burde, B. de la and Choate, M. S. 1972: Does asymptomatic lead exposure in children have latent sequelae? *J. Pediar.*, 81, 1088–91.

Burton, I., Billingsley, D., Blacksell, M., Chapman, V., Kirkby, A. V., Foster, L. and Wall, G. 1974: Public response to a successful air pollution control programme. In J. A. Taylor (ed.), *Climatic Resources and Economic Activity*, Newton Abbot: David and Charles.

Butzer, K. W. 1980: Adaptation to global environmental change. *Prof. Geog.*, 32, 269–78.

Campbell, W. A. and Heath, M. S. 1977: Air pollution legislation and regulations. In A. C. Stern (ed.), *Air Pollution. Vol. 5*, Air Quality Management, 374–7.

Carter, F. W. 1985: Pollution problems in post-war Czechoslovakia. *Trans. Inst. Brit. Geogr. New Series*, 10, 17–44.

Cass, G. R. 1979: On the relationship between sulphate air quality and visibility with examples in Los Angeles. *Atmospheric Environ.*, 13, 1069–84.

Chamberlain, A. C. 1959: Deposition of iodine-131 in northern England in October 1957. *Quat. J. Royal Met. Soc.*, 85, 350–61.

Chandler, T. J. 1961: Surface breeze effects of Leicester's heat-island. *East Midland Geographer*, 2, 32–8.

Chandler, T. J. 1965: *The Climate of London*. London: Hutchinson.

Chandler, T. J. 1976: Urban climatology and its relevance to urban design. *WMO Technical Note*, 149, Geneva: World Meteorological Organization.

Chang, S. -D. 1985: Urban environmental quality in China: a luxury or a necessity? *China Geographer*, 12, 81–99.

Changnon, S. A. 1968: The La Porte anomaly – fact or fiction? *Bull. Amer. Met. Soc.*, 49, 4–11.

Changnon, S. A. 1970: Recent studies of urban effects on precipitation in the United States. *WMO Technical Note*, 108, Geneva: World Meteorological Organization, 327–43.

Changnon, S. A. 1978: Urban effects on severe local storms at St Louis. *J. App. Meteor.*, 17, 578–86.

Changnon, S. A. 1980: More on the La Porte anomaly: a review. *Bull. Amer. Met. Soc.*, 61, 702–11.

Changnon, S. A. Huff, F. A., Schickedanz, P. T. and Vogel, J. L. 1977: *Summary of METROMEX, Volume 1: Weather Anomalies and Impacts*. Urbana: Illinois State Water Survey Division, Bulletin No. 62.

Chase, D. J. 1973: Eco-journalism and the failure of crisis reporting. *J. Environ. Education*, 5, 4–7.

Cherfas, J. 1980: US acts to clean up its act. *New Scientist*, 85, 164–6.

Chock, D. P., Kumar, S. and Herrmann, R. W. 1982: An analysis of trends in oxidant air quality in the South Coast Air Basin of California. *Atmospheric Environ.*, 16, 2615–24.

Clark, W. C., Cook, K. H., Marland, G., Weinberg, A. M., Rotty, R. M., Bell, P. R., Allison, L. J. and Cooper, C. L. 1982: The carbon dioxide question:

perspectives for 1982. In W. C. Clark (ed.), *Carbon Dioxide Review 1982*, Oxford: Oxford University Press, 3–44.

Clarke, R. and Timberlake, L. 1982: *Stockholm Plus Ten. Promises, Promises? The Decade Since the 1972 UN Environment Conference*. London: Earthscan.

Cogbill, C. V. and Likens, G. E. 1974: Acid precipitation in the NE USA. *Water Resources Res.*, 10, 1133–7.

Coggle, J. E. 1983: *Biological Effects of Radiation*. 2nd edition. London: Taylor & Francis.

Commins, B. T. and Waller, R. E. 1967: Observations from a ten year study of pollution at a site in the city of London. *Atmospheric Environ.*, 1, 49–68.

Commission of the European Communities, Joint Research Centre 1982: Isotopic lead experiment status report. *Commission of the European Communities Report*, No. EUR 8352 EN, Luxembourg.

Cooper, C. F. 1982: Food and fiber. In W. C. Clark (ed.) *Carbon Dioxide Review 1982*, Oxford: Oxford University Press, 299–320.

Copley International Corporation 1970: *National Survey of the Odor Problem*. Springfield, Virginia: US Dept. of Commerce, National Technical Information Service, Publ. No. PB-194 376.

Copley International Corporation 1971: *A Study of the Social and Economic Impact of Odors, Phase II*. Springfield, Virginia: US Dept. of Commerce, National Technical Information Service, Publ. No. PB-205 936.

Costle, D. M. 1979: Dollars and sense: putting a price tag on pollution. *Environment*, 21, 25–7.

Covey, C., Schneider, S. H. and Thompson, S. L. 1984: Global atmospheric effects of massive smoke injections from a nuclear war: results from general circulation model simulations. *Nature*, 308, 21–5.

Cox, R. A., Eggleton, A. E. J., Derwent, R. G., Lovelock, J. E. and Pack, D. H. 1975: Long-range transport of photochemical ozone in north-western Europe. *Nature*, 255, 118–21.

Creech Jr, J. L. and Johnson, M. N. 1974: Angiosarcoma of liver in the manufacture of polvinyl chloride. *J. Occup. Med.*, 16, 150–1.

Crick, M. J. and Linsley, G. S. 1983: An assessment of the radiological impact of the Windscale Reactor fire 1957. *National Radiological Protection Board Report*, 135, London: HMSO.

Crossland, J. 1978: Reporting pollution. *Environment*, 20, 29–31.

Crutzen, P. J. 1972: SSTs – a threat to the Earth's ozone shield. *Ambio*, 1, 41–51.

Crutzen, P. J. 1974: Estimates of possible future ozone reductions from continued use of fluoro-chloro-methanes (CF2CL2, CFCL3). *Geophys. Res. Letters*, 1, 205–8.

Crutzen, P. J. 1976: Upper limits on atmospheric ozone reductions following increased applications of fixed nitrogen to the soil. *Geophys. Res. Letters*, 3, 169–72.

Crutzen, P. J. and Arnold, F. 1986: Nitric acid cloud formation in the cold Antarctic stratosphere: a major cause for the springtime 'ozone-hole'. *Nature*, 324, 651–5.

Crutzen, P. J. and Birks, J. W. 1982: The atmosphere after a nuclear war: twilight at noon. *Ambio*, 11, 114–25.

Daisey, J. M., Kneip, T. J., Wang, M-x., Ren, L-x., and Lu, W-x 1983: Organic and elemental composition of airborne particulate matter in Beijing, Spring 1981. *Aerosol Science and Technology*, 2, 407–15.

David, O. J., Hoffman, S. P., Clark, J., Grad, G. and Sverd, J. 1983: The relationship of hyperactivity to moderately elevated lead levels. *Arch. Environ. Health*, 38, 341–6.

Davidson, A. 1986: Comment on 'ten-year ozone trends in California and Texas'. *J. Air Pollut. Control Ass.*, 36, 597–8.

Davies, T. D., Abrahams, P. W., Tranter, M., Blackwood, I., Brimblecombe, P. and Vincent, C. E. 1984: Black acidic snow in the remote Scottish Highlands. *Nature*, 312, 58–61.

DeLucia, J. A. and Adams, W. C. 1977: Effects of ozone inhalation during exercise on pulmonary function and blood biochemistry. *J. Applied Physiology*, 43, 75–81.

Derwent, R. G. 1982: Two-dimensional model studies of the impact of aircraft exhaust emissions on tropospheric ozone. *Atmospheric Environ.*, 16, 1997–2007.

Derwent, R. G. and Stewart, H. N. M. 1973: Air pollution from the oxides of nitrogen in the UK *Atmospheric Environ.*, 7, 385–401.

Derwent, R. G., McInnes, G., Stewart, H. N. M. and Williams, M. L. (1976) *The Occurrence and Significance of Air Pollution by Photochemically Produced Oxidant in the British Isles 1972–75*. Stevenage: Warren Spring Laboratory, Report LR 227 (AP).

Dienes, L. 1974: Environmental disruption in Eastern Europe. In I. Volgyes (ed.) *Environmental Deterioration in the Soviet Union and Eastern Europe*, New York: Praeger, 141–58.

Downs, A. 1972: Up and down with ecology – the 'issue attention cycle'. *Public Interest*, 28, 38–51.

Dworkin, J. M. & Pijawka, K. D. 1981: Air quality and perception: explaining change in Toronto, Ontario, *Working Paper, Inst. of Environ. Studies, University of Toronto*, EPR-9.

Eaglemann, R. J. 1981: Focus on the ozone layer. *Environ. Conservation*, 8, 147–9.

Economic Commission for Europe 1985: *Air Pollution Across Boundaries*. New York: United Nations, Air Pollution Studies 2.

Eggleton, A. E. J. 1969: The chemical composition of atmospheric aerosols on Tee-side and its relation to visibility. *Atmospheric Environ.*, 3, 355–72.

Ehrlich, P. and nineteen others 1983: Long-term biological consequences of nuclear war. *Science*, 222, 1293–1300.

Einbender, G., Bakalian, A., Wall, T., Hoagland, P. and Kamlet, K. S. 1982: The case for immediate controls on acid rain. *Materials and Society*, 6, 251–82.

Eisenbud, M. 1973: *Environmental Radioactivity*. 2nd edition. London and New York: Academic Press.

El-Hinnawi, E. E. 1980: *Nuclear Energy and the Environment*. Oxford: Pergamon

Press, United Nations Program, Nairobi, Kenya.

Eliassen, A. and Saltbones, J. 1983: Modelling of long-range transport of sulphur over Europe: a two-year model run and some model experiments. *Atmospheric Environ.*, 17, 1457–73.

Elsom, D. M. 1978a: The changing nature of a meteorological hazard. *J. Meteorology, UK*, 3, 297–9.

Elsom, D. M. 1978b: Meteorological aspects of air pollution episodes in urban areas. *Discussion Paper in Geography, Oxford Polytechnic*, 5.

Elsom, D. M. 1979: Air pollution episode in Greater Manchester. *Weather*, 34, 277–86.

Elsom, D. M. 1982: Drastic reduction in the United Kingdom Air Pollution Monitoring Network. *Geography*, 67, 134–7.

Elsom, D. M. 1983: Pollution. In M. Pacione (ed.), *Progress in Urban Geography*, Beckenham: Croom Helm, 251–77.

Elsom, D. M. 1984a: Los Angeles smog threatens 1984 Olympic Games. *Weather*, 39, 200–7.

Elsom, D. M. 1984b: The Olympics under a cloud. *Geographical Mag.*, 56, 338–40.

Elsom, D. M. 1984c: Climatic change induced by a large-scale nuclear exchange. *Weather*, 39, 268–71.

Elsom, D. M. 1985: Climatological effects of a nuclear exchange: a review. In D. M. Pepper and A. Jenkins (eds), *The Geography of Peace and War*, Oxford: Basil Blackwell.

Elsom, D. M. and Meaden, G. T. 1982: Suppression and dissipation of weak tornadoes in metropolitan areas: a case-study of Greater London. *Mon. Weath. Rev.*, 110, 745–56.

Engen, T. 1972: Use of sense of smell in determining environmental quality. In W. A. Thomas (ed.), *Indicators of Environmental Quality*, New York: Plenum Press.

Engstrom, A. 1972: *Sweden's Case Study for the United Nations Conference on the Human Environment, 1972: Air Pollution across National Boundaries. The Impact on the Environment of Sulphur in Air and Precipitation*. Stockholm: Royal Ministry for Foreign Affairs and Royal Ministry of Agriculture.

Enloe, C. H. 1975: *The Politics of Pollution in a Comparative Perspective: Ecology and Power in Four Nations*. New York: David McKay.

Environmental Resources Limited 1983: *Acid Rain, A Review of the Phenomenon in the EEC and Europe*. London: Environmental Resources Limited.

Faith, W. L. and Atkisson Jr, A. A. 1972: *Air Pollution*. 2nd edition. New York: Wiley Interscience.

Farman, J. C., Gardiner, B. G. and Shanklin, J. D. 1985: Large losses of total ozone in Antarctica reveal seasonal C1Ox/NOx interaction. *Nature*, 315, 207–10.

Farquhar, J. T. 1983: The policies of the European Community towards the environment – the 'dangerous substances' Directive. *J. Planning & Environ. Law*, March, 145–55.

Findlater, J. 1974: The low-level cross-equatorial air current of the western Indian Ocean during the northern summer. *Weather*, 29, 411–16.

Findlay, B. F. and Hirt, M. S. 1969: An urban-induced meso-circulation. *Atmospheric Environ.*, 3, 537–542.

Fishman, J., Ramanathan, V., Crutzen, P. J. and Liu, S. C. 1979: Tropospheric ozone and climate. *Nature*, 282, 818–20.

Flesh, R. D., Burns, J. C. and Turk, A. 1974: An evaluation of community problems caused by industrial odors. In A. Turk, J. W. Johnston Jr, and D. G. Moulton (ed.), *Human Responses to Environmental Odors*, New York: Academic Press, 33–44.

Flohn, H. 1977: Climate and energy: a scenario to a 21st century problem. *Climatic Change*, 1, 5–20.

Flower, R. J. and Battarbee, R. W. 1983: Diatom evidence for recent acidification of two Scottish lochs. *Nature*, 305, 130–3.

Folinsbee, L. J. Silverman, F. and Shephard, R. J. 1975: Exercise responses following ozone exposure. *J. Applied Physiology*, 38, 996–1001.

Folinsbee, L. J. Wagner, J. A., Borgia, J. F., Drinkwater, B. L. Gliner, J. A. and Bedi, J. F. 1978: *Environmental Stress: Individual Human Adaptations*. New York: Academic Press.

Fowler, D., Cape, J. N., Leith, I. D., Paterson, I. S., Kinnaird, J. W. and Nicholson, I. A. 1982: Rainfall acidity in northern Britain. *Nature*, 297, 383–6.

Fujita, T. T. 1973: Tornadoes around the world. *Weatherwise*, 26, 56–62, 78–83.

Fuller, J. G. 1977: *The Poison that Fell From the Sky*. New York: Random House.

Galloway, J. N., Likens, G. E. and Edgerton, E. S. 1976: Acid precipitation in the Northeastern US: pH and acidity. *Science*, 194, 722–4.

GEMS (Global Environment Monitoring System) 1984: *Urban Air Pollution 1973–1980*. Geneva: World Health Organization.

Gilbert, O. L. 1974: An air pollution survey by school children. *Environ. Pollution*, 6, 170–80.

Giles, J. W. 1984: The Canadian acid rain program. In T. C. Elliott and R. G. Schwieger (eds), *The Acid Rain Sourcebook*, New York: McGraw-Hill, 56–63.

Gilliland, R. L. 1982: Solar, volcanic, and carbon dioxide forcing of recent climatic changes. *Climatic Change*, 4, 111–31.

Gilliland, R. L. and Schneider, S. H. 1984: Volcanic, carbon dioxide and solar forcing of Northern and Southern Hemisphere surface air temperature. *Nature*, 310, 38–41.

Glass, R. I. 1975: A perspective on environmental health in the USSR. *Arch. Environ. Health*, 30, 391–5.

Gold, E. 1956: Smog. The rate of influx of surrounding cleaner air. *Weather*, 11, 230–2.

Goldman, M. I. 1970: The convergence of environmental disruption. *Science*, 170, 37–42.

Goldsmith, J. R. 1986: *Environmental Epidemiology. Epidemiological Investigations of Community Environmental Health Problems.* Boca Raton, Florida: CRC Press.

Goldsmith, P., Tuck, A. F., Foot, J. S., Simmons, E. L. and Newson, R. L. 1973: Nitrogen oxides, nuclear weapon testing, Concorde and stratospheric ozone. *Nature*, 244, 545–51.

Goodenough, R. 1983: Recent developments in California's environmental review process – an evaluation of the use of environmental impact reports. *Intern. J. Environ. Studies*, 20, 181–9.

Gorham, E. 1982: What to do about acid rain. *Tech. Review*, 85, 59–70.

Green, L. Jr. 1984: Coal cleaning: the first step. In T. C. Elliott and R. G. Schwieger (eds), *The Acid Rain Sourcebook*, New York: McGraw-Hill, 112–22.

Greenburg, L., Field, F., Reed, J. I., and Erhard, C. I. 1962: Air pollution and morbidity in New York City. *J. American Medical Society*, 182, 161–4.

Gregory, R. 1972: Conservation, planning and politics: some aspects of the contemporary British scene. *Intern. J. Environ. Studies*, 4, 33–9.

Grennfelt, P. and Schjoldager, J. 1984: Photochemical oxidants in the troposphere: a mounting menace. *Ambio*, 13, 61–7.

Gross, M. 1982: Computer simulation in urban planning and air pollution control. *J. Environ. Systems*, 11, 257–69.

Groves, K. S. and Tuck, A. F. 1979: Simultaneous effects of carbon dioxide and chlorofluromethanes on stratospheric ozone. *Nature*, 280, 127–9.

Gunningham, M. 1974: *Pollution, Social Interest and the Law.* London: Martin Robertson.

Guo, H. 1981: Environmental protection in China. *Beijing Review*, 26, 12–15.

Hackney, J. D., Linn, W. S., Mohler, J. G. and Colier, C. 1977a: Adaptation to short-term respiratory effects of ozone in men exposed repeatedly. *J. Applied Physiology*, 43, 82–5.

Hackney, J. D., Linn, W. S., Law, D. C., Karuza, S. K., Greenburg, H., Buckley, R. D. and Pedersen, E. E. 1975: Experimental studies on human health effects of air pollutants III. Two-hour exposure to ozone alone and in combination with other pollutant gases. *Arch. Environ. Health*, 30, 385–90.

Hackney, J. D., Linn, W. S., Karuza, S. K., Buckley, R. D., Law, D. C., Bates, D. V., Hazucha, M., Pengelly, L. D. and Silverman, F. 1977b: Effects of ozone exposure in Canadians and southern Californians. *Arch. Environ. Health*, 32, 110–16.

Haigh, N. 1984: *EC Environmental Policy and Britain.* London: Environmental Data Services.

Hall, L. W. 1977: Environmental management in China: a two-part perspective. Part II: the costs and benefits of mobilising the masses against waste. *China Geographer*, 8, 37–56.

Hammer, D. I., Hasselblad, V., Portnoy, B. and Wehrle, P. F. 1974: Los Angeles student nurse study: daily symptom reporting and photochemical oxidants. *Arch. Environ. Health*, 28, 255–60.

Handscombe, C. and Elsom, D. M. 1982: Rationalisation of the National Survey of Air Pollution Monitoring Network of the United Kingdom using spatial correlation analysis: a case-study of the Greater London area. *Atmospheric Environ.*, 16, 1061–70.

Hanna, S. R. and Gifford, F. A. 1975: Meteorological effects of energy dissipation at large power parks. *Bull. Amer. Met. Soc.*, 56, 1069–76.

Hansen, J., Johnson, D., Lacis, A., Lebedeff, S., Lee, P., Rind, D. and Russell, G. 1981: Climate impact of increasing atmospheric carbon dioxide. *Science*, 213, 957–66.

Hansen, J., Johnson, D., Lacis, A., Lebedeff, S., Lee, P., Rind, D. and Russell, G. 1983: Climatic effects of atmospheric carbon dioxide. *Science*, 220, 873–5.

Hare, F. K. 1982: The carbon dioxide question – threat or opportunity? *Environ. Conservation*, 9, 343–4.

Harris, B. D. and Smith, K. 1982: Cleaner air improves visibility in Glasgow. *Geography*, 67, 137–9.

Harris, G. R. 1981: Positive impacts of environmental policy on business in the United States. *Intern. J. Environ. Studies*, 16, 75–83.

Harrison, H. 1970: Stratospheric ozone with added water vapour: influence of high-altitude aircraft. *Science*, 170, 734–6.

Harrison, R. and McGoldrick, B. 1981: Mapping artificial heat release in Great Britain. *Atmospheric Environ.*, 15, 667–74.

Hart, G. 1982: Clean air: time for responsible reform. *J. Air Pollut. Control Ass.*, 32, 14–18.

Harte, J. 1983: An investigation of acid precipitation in Qinghai Province, China. *Atmospheric Environ.*, 17, 403–8.

Hawkes, N., Lean, G., Leigh, D., McKie, R., Pringle, P. and Wilson, A. 1986: *The Worst Accident in the World*. London: Pan Books/Heinemann.

Hawksworth, D. L. and Rose, F. 1970: Qualitative scale for estimating sulphur dioxide air pollution in England and Wales using epiphytic lichens. *Nature*, 227, 145–8.

Hawksworth, D. L. and Rose, F. 1976: *Lichens as Pollution Monitors*. London: Edward Arnold, Inst. of Biology, Studies in Biology No. 66.

Hazucha, M., Silverman, F., Parent, C., Field, S. and Bates, D. V. 1973: Pulmonary function in man after short-term exposure to ozone. *Arch. Environ. Health*, 27, 183–8.

Hazucha, M. J., Ginsberg, J. F., McDonnell, W. F., Haak Jr, E. D., Pimmel, R. L., Salaam, S. A., House, D. E. and Bromberg, P. A. 1983: Effects of 0.1 ppm nitrogen dioxide on airways of normal and asthmatic subjects. *J. Applied Physiology*, 54, 730–9.

Heath, D. F. 1988: Non-seasonal changes in total column ozone from satellite observations, 1970–86. *Nature*, 332, 219–27.

Heck, W. W., Taylor, O. C., Adams, R., Bingham, G., Miller, J., Preston, E. and Weinstein, L. 1982: Assessment of crop loss from ozone. *J. Air Pollut. Control Assoc.*, 32, 353–8.

Heck, W. W., Adams, R. M., Cure, W. W., Heagle, A. S., Heggerstad, H. E., Kohut, R. J., Kress, L. W., Rawlings, J. O. and Taylor, O. C. 1983: A reassessment of crop loss from ozone. *Environ. Sci. Technology*, 17, 572–81.

Heicklin, J. 1981: Control of photochemical smog by diethylhydroxylamine. *Atmospheric Environ.*, 15, 229–41.

Hendrey, G. R., Baalsrud, K., Traaen, T. S., Laake, M. and Raddum, G. 1976: Some hydrological changes. *Ambio*, 5, 224–7.

Hibbard, W. R. 1982: Acid precipitation: a critique of present knowledge and proposed action. *Materials and Society*, 6, 357–83.

Hidy, G. M., Mueller, P. K. and Tong, E. Y. 1978: Spatial and temporal distributions of airborne sulfate in parts of the US *Atmospheric Environ.*, 12, 735–52.

Hidy, G. M., Hansen, D. A., Henry, R. C., Ganesan, K. and Collins, J. 1984: Trends in historical acid precursor emissions and their airborne and precipitation products. *J. Air Pollut. Control Ass.*, 31, 333–54.

Hinrichsen, D. 1986: Multiple pollutants and forest decline. *Ambio*, 15, 258–65.

Hogstrom, U. 1978: Meteorology of cities as related to air pollution dispersion. WMO Symposium on Boundary Layer Physics applied to specific problems of air pollution. *WMO Technical Note*, 510, Geneva: World Meteorological Organization. 53–61.

Hohenemser, C., Deicher, M., Ernst, A., Hofsass, H., Lindner, G. and Recknagel, E. 1986: Chernobyl: an early report. *Environment*, 28, 6–13, 30–43.

Holdgate, M. W., Kassas, M. and White, G. H. 1982a: *The World Environment 1972–1982*. Dublin: Tycooly International.

Holdgate, M. W., Kassas, M. and White, G. F. 1982b: World environmental trends between 1972 and 1982. *Environ. Conservation*, 9, 11–29.

Hosler, C. L. and Landsberg, H. E. 1977: The effect of localised man-made heat and moisture sources in mesoscale weather modification. In National Academy of Sciences, *Energy and Climate*, Washington, D.C.: Studies in Geophysics, Geophysics Study Committee, 96–105.

Hov, O. 1984: Ozone in the troposphere: high level pollution. *Ambio*, 13, 73–9.

Huff, F. A. and Changnon, S. A. 1973: Precipitation modification by major urban areas. *Bull. Amer. Met. Soc.*, 54, 1220–32.

Hulm, P. 1982: WMO reiterates risk of ozone depletion. *Ambio*, 11, 70.

Huntzicker, J. J., Friedlander, S. K. and Davidson, C. 1975: Material balance for automobile-emitted lead in Los Angeles basin. *Environ. Sci. Technology*, 9, 448–57.

ICRP-26 (International Commission on Radiological Protection) 1977: Recommendations of the International Committee on Radiological Protection. *Annals of the ICRP*, 1, 1–53.

ICRP-29 (International Committee on Radiological Protection) 1979: Radionuclide release into the environment: assessment of doses to man. *Annals of the ICRP*, 2, 1–76.

Idso, S. B. 1980: The climatological significance of a doubling of Earth's atmospheric carbon dioxide concentration. *Science*, 207, 1462–3.

Idso, S. B. 1984: A review of reports dealing with the greenhouse effect of atmospheric carbon dioxide. *J. Air Pollut. Control Ass.*, 34, 553–5.

International Atomic Energy Agency 1986: *Nuclear Power Reactors in the World*. Vienna: International Atomic Energy Agency.

Izmerov, N. F. 1973: Control of air pollution in the USSR. *Public Health Paper, World Health Organization, Geneva*, 54.

Jacobson, H. K. & Kay, D. A. 1983: A framework for analysis. In D. A. Kay, & H. K. Jacobson (eds), *Environmental Protection*, Allanheld: Osmun Publishers, 1–21.

Jernelov, A. 1983: Acid rain and sulphur dioxide emissions in China. *Ambio*, 12, 362.

Johnson, A. H. and Siccama, T. G. 1983: Acid deposition and forest decline. *Environ. Sci. Technology*, 17, 249A–305A.

Johnson, S. P. 1979: *The Pollution Control Policy of the European Communities*. London: Graham and Trotman.

Johnston, H. 1971: Reduction of stratospheric ozone by nitrogen oxide catalysts from supersonic transport exhaust. *Science*, 173, 517–22.

Jones, C. O. 1973: Air pollution and contemporary environmental politics. *Growth and Change*, 4, 22–7.

Kagawa, J. and Toyama, T. 1975: Photochemical air pollution; its effects on respiratory functions of elementary school children. *Arch. Environ. Health*, 30, 117–22.

Kallend, A. S., Marsh, A. R. W., Pickles, J. H. and Proctor, M. V. 1983: Acidity of rain in Europe. *Atmospheric Environ.*, 17, 127–37.

Kates, R. W. 1977: Risk assessment of environmental hazard. *Scope Report*, 8, Chichester: Wiley.

Kazis, R. and Grossman, R. L. 1982: Job-taker or job-maker? *Environment*, 24, 13–20, 43.

Keating, G. M. 1978: Relation between monthly variations of global ozone and solar activity. *Nature*, 274, 873–4.

Keith, R. W. 1980: *A Climatological-Air Quality Profile: California South Coast Air Basin*. El Monte, California: South Coast Air Quality Management District, Air Programs Division Report.

Kelley, D. R. 1980: East-West environmental cooperation. *Environment*, 22, 29–37.

Kelley, D. R., Stunkel, K. R. and Westcott, R. R. 1976: *The Economic Superpowers and the Environment*. San Francisco: Freeman, 64–70.

Kellogg, W. W. 1978: Global influences of mankind on the climate. In J. Gribbin (ed.), *Climatic Change*, Cambridge: Cambridge University Press, 205–27.

Kellogg, W. W. and Schware, R. 1981: *Climatic Change and Society: consequences of increasing atmospheric carbon dioxide*. Boulder: Westview Press.

Kelly, J. J. 1979: Innovations in air pollution abatement technology. In *Proc. of Seminar: Air Pollution – Impacts and Control*, Dublin: National Board for Science and Technology, 70–86.

Kemf, E. 1984: Air pollution in the Arctic. *Ambio*, 13, 122–3.

Kerr, H. D., Kulle, T. J., McIlhany, M. L. and Swidersky, P. 1975: Effects of ozone on pulmonary function in normal subjects. An environmental-chamber study. *Amer. Rev. Resp. Dis.*, 111, 763–73.

Kerr, H. D., Kulle, T. J., McIlhany, M. L. and Swidersky, P 1978: Effects of nitrogen dioxide on pulmonary function in human subjects: an environmental chamber study. *US Environmental Protection Agency, Washington, D.C., Report*, EPA-600/1-78-025.

Khalil, M. A. K. and Rasmussen, R. A. 1984: Carbon monoxide in the Earth's atmosphere: increasing trend. *Science*, 224, 54–6.

Kimball, B. A. 1982: *Report 11* . Phoenix, Arizona: US Water Conservation Laboratory.

Kinzelbach, W. K. H. 1982: Energy and environment in China. *Environ. Policy and Law*, 8, 78–82.

Kinzelbach, W. K. H. 1983: China: energy and environment. *Environ. Management*, 7(4), 303–10.

Kneese, A. V. 1984: *Measuring the Benefits of Clean Air and Water*. Washington, D.C.: Resources for the Future.

Ko, N. W. M. 1978: Traffic noise in a high rise city. *Applied Acoustics*, 11, 225–39.

Koenig, J. Q., Pierson, W. E. and Frank, R. 1980: Acute effects of inhaled sulphur dioxide plus sodium chloride aerosol on pulmonary function in asthmatic adolescents. *Environ. Res.*, 22, 145.

Koenig, J. Q., Pierson, W. E. and Frawke, R. 1979: Acute effects of inhaled sulphur dioxide and exercise on pulmonary function in asthmatic adolescents. *J. Allergy and Clinical Immunology*, 63, 154.

Komarov, B. 1980a: *The Destruction of Nature in the Soviet Union*, Nottingham: Pluto Press.

Komarov, B. 1980b: Soviet conservation: a bear with no claws. *New Scientist*, 88, 514–5.

Komhyr, W. D., Barrett, E. W., Slocum, G. and Weickmann, H. K. 1971: Atmospheric total ozone increase during the 1960s. *Nature*, 232, 390–1.

Kondratyev, K. Ya. and Nikolsky, G. A. 1979: The stratospheric mechanism of solar and anthropogenic influences on climate. In B. M. McCormac and T. A. Seliga (eds), *Solar-Terrestrial Influences on Weather and Climate*, Dordrecht: Reidel, 317–22.

Kormondy, E. J. 1980: Environmental protection in Hungary and Poland *Environment*, 22, 31–7.

Krier, J. E. and Ursin, E. 1977: *Pollution and Policy: a case essay on Californian and federal experience with motor vehicle air pollution 1940–75*. Berkeley and Los Angeles: University of California Press.

Lal, M., Dube, S. K. Sinha, P. C. and Jain, A. K. 1986: Potential climatic consequences of increasing anthropogenic constituents in the atmosphere. *Atmospheric Environ.*, 20, 639–42.

Landsberg, H. E. 1981: *The Urban Climate*. New York: Academic Press.

Lansdown, R., Yule, W., Urbanowicz, M. -A. and Miller, I. B. 1983: Blood lead, intelligence, attainment and behaviour in school children: overview of a pilot study. In M. Rutter and R. Russell Jones (eds), *Lead Versus Health: sources and effects of low level lead exposure*, Chichester: Wiley.

Last, J. A., Greenburg, D. B. and Castleman, W. L. 1979: Ozone-induced alterations in collagen metabolism of rat lungs. *Toxicology and Applied Pharmocology*, 51, 247–58.

Laszlo, E. 1984: The state of the environment in Hungary. *Ambio*, 13, 93–108.

Lave, L. B. and Omenn, G. S. 1981: *Clearing the Air: reforming the Clean Air Act*. Washington, D. C.: Brookings Institution.

Lave, L. B. and Seskin, E. 1977: *Air Pollution and Human Health*. Baltimore: Johns Hopkins University Press.

Lawther, P. J., Macfarlane, A. J. Waller, R. W. and Brooks, A. G. F. 1975: Pulmonary function and sulphur dioxide: some preliminary findings. *Environ. Res.*, 10, 355–69.

Laxen, D. 1985: Nitrogen dioxide: an air quality problem in London? *London Environ. Bulletin*, 3, 10–12.

Lee, D. O. 1983a: Trends in summer visibility in London and southern England 1962–1979. *Atmospheric Environ.*, 17, 151–9.

Lee, D. O. 1983b: Pollution over London. *Geographical Mag.*, 55, 278–9.

Lee, D. O. 1985: Britain's imported air pollution. *Geography*, 70, 257–8.

Lee, R. E. Jr., Caldwell, J. S. and Morgan, G. B. 1972: The evaluation of methods for measuring suspended particulates in air. *Atmospheric Environ.*, 6, 593–622.

Lee, S. D. 1980: *Nitrogen Oxides and Their Effects on Health*. Ann Arbor: Ann Arbor Science.

Leonardos, G. 1974: A critical review of regulations for the control of odors. *J. Air Pollut. Control Ass.*, 24, 456–68.

Levitt, R. 1980: *Implementing Public Policy*. London: Croom Helm.

Lewis, D. and Davis, W. 1986: *Joint Report of the Special Envoys on Acid Rain*. Ontario: Environment Canada.

Lewis, W. M. and Grant, M. C. 1980: Acid precipitation in the western United States. *Science*, 207, 176–7.

Li, W. 1979: Trees that help to reduce pollution. *Clean Air*, 9(1), 6–7.

Lie, S. C., Cicerone, R. J., Donahue, T. M. and Chameides, W. L. 1976: Limitation of fertiliser induced ozone reduction by the long lifetime of the reservoir of fixed nitrogen. *Geophys. Res. Letters.*, 3, 157–60.

Likens, G. E. and Bormann, F. H. 1974: Acid rain; a serious regional environmental problem. *Science*, 184, 1176–79.

Likens, G. E. and Butler, T. J. 1981: Recent acidification of precipitation in North America. *Atmospheric Environ.*, 15, 1103–9.

Likens, G.E. Bormann, F. H. and Johnson, N. M. 1972: Acid rain. *Environment*, 14, 33–40.

Likens, G. E., Wright, R. F., Galloway, J. N. and Butler, T. J. 1979: Acid rain. *Scientific American*, 241, 39–47.

Lin, G. -Y. 1981: Simple Markov chain model of smog probabilities in the South Coast Air Basin of California. *Prof. Geog.*, 33, 228–36.

Lin, G. -Y. and Bland, W. R. 1980: Spatio-temporal variations in photochemical smog concentrations in Los Angeles County. *California Geographer*, 20, 28–52.

Linn, W. S., Jones, M. P., Bachmayer, E. A. Spier, C. E., Mazur, S. F., Avol, E. L. and Hackney, J. D. 1980: Short-term respiratory effects of polluted ambient air: a laboratory study of volunteers in a high-oxidant community. *Amer. Review of Resp. Dis.*, 121, 243–9.

Liroff, R. A. 1980: *Air Pollution Offsets: trading, selling, and banking.* Washington, D. C.: Conservation Foundation.

Liss, P. S. and Crane, A. J. 1983: *Man-made Carbon Dioxide and Climatic Change.* Norwich: Geo Books.

Lovins, A. B. and Lovins, L. H. 1982: Energy, economics and climate – an editorial. *Climatic Change*, 4, 217–20.

Lubinska, A. 1984: Towards lead-free auto fuel. *Nature*, 311, 401.

Lundqvist, L. J. 1980: *The Hare and the Tortoise: Clean Air Policies in the United States and Sweden.* Ann Arbor: University of Michigan Press.

Lyons, W. A. and Dooley Jr, J. C. 1978: Satellite detection of long-range pollution transport and sulphate aerosol hazes. *Atmospheric Environ.*, 12, 621–31.

McAfee, J. 1982: Clean air, energy, and jobs: can we have them all? *J. Air Pollut. Control Ass.*, 32, 8–18.

McCafferty, W. B. (1981) *Air Pollution and Athletic Performance.* Springfield, Illinois: Charles C. Thomas.

McCormick, J. (1985) *Acid Earth: The Global Threat of Acid Precipitation.* London: International Institute for Environment and Development (Earthscan).

MacDonald, G. J. 1982: *The Long-term Impacts of Increasing Atmospheric Carbon Dioxide Levels.* Cambridge, Mass.: Ballinger.

McDonald, P. I. 1975: New policies for control of environmental pollution in China. *Solid Wastes*, 65(11), 543–5.

McElroy, M. B., Elkins, J. W., Wofsy, S. C. and Yung, Y. L. 1976: Sources and sinks for atmospheric nitrous oxide. *Rev. Geophys. Space Phys.*, 14, 143–50.

Macfarlane, A. 1978: Daily mortality and environment in English conurbations. II. Deaths during summer hot spells in Greater London. *Environ.Res.*, 15, 322–41.

McGinty, L. 1977: Air pollution underestimated in London. *New Scientist*, 75, 141.

McGoldrick, B. 1980: Artificial heat release from Greater London, 1971. *Physics Division Energy Workshop Report, Dept, of Physical Sciences, Sunderland Polytechnic*, 20.

Magat, W. A. 1982: *Reform of Environmental Regulation.* Cambridge, Mass.: Ballinger.

Marchetti, C. 1977: On geoengineering and the carbon dioxide problem. *Climatic Change*, 1, 59–68.

Maracek, J., Shapiro, I. M., Burke, A., Katz, S. H. and Hediger, M. L. 1983: Low-level lead exposure in childhood influences neuropsychological performance. *Arch. Environ. Health*, 38, 355–9.

Marshall, E. 1978: EPA smog standard attacked by industry, science advisers. *Science*, 202, 949–50.

Marshall, E. 1979: Smog's not so bad, EPA decides. *Science*, 203, 529.

Martin, A. E. 1964: Mortality and morbidity statistics in air pollution. *Proc. Roy. Soc. Med.*, 57, 969–75.

Martin, A. E. and Bradley, W. H. 1960: Mortality, fog and atmospheric pollution - an investigation during the winter of 1958–59. *Mon. Bull. Min. Health, Public Health Lab. Serv.*, 19, 56–72.

Martin, W. 1975: Legislative air pollution strategies in various countries. *Clean Air*, 9, 28–32.

Masters, B. R. 1974: The City of London and clean air 1273 A. D. to 1973 A. D. *Clean Air*, 4, 22–8.

Matthews, W. H., Kellogg, W. W. and Robinson, G. D. 1971: *Inadvertent Climate Modification, Study of Man's Impact on Climate (SMIC)*, Cambridge, Mass: MIT Press.

Maugh, T. H. 1984: What is the risk from chlorofluorocarbons? *Science*, 223, 1051–2.

Mazumdar, S. and Sussman, N. 1983: Relationships of air pollution to health: results from the Pittsburgh study. *Arch. Environ. Health*, 38, 17–24.

Mazumdar, S., Schimmel, H. and Higgins, I. T. T. 1982: Relation of daily mortality to air pollution: an analysis of 14 London winters, 1958/59–1971/72. *Arch. Environ. Health*, 37, 213–20.

Medvedev, Z. 1979: *Nuclear Disaster in the Urals*. New York: Norton.

Meetham, A. R., Bottom, D. W., Cayton, S., Henderson-Sellers, A. and Chambers, D. 1981: *Atmospheric Pollution: its history, origins and prevention*. Oxford: Pergamon.

Melia, R. J. W., Florey, C. du V., Altman, D. S. and Swan, A. V. 1977: Association between gas cooking and respiratory disease in children. *Brit. Med. J.*, 2, 149–52.

Melia, R. J. W., Florey, C. du V., Darby, S. C., Palms, E. D. and Goldstein, B. D. 1978: Differences in nitrogen dioxide levels in kitchens with gas or electric cookers. *Atmospheric Environ.*, 12, 1379–81.

Mikami, R. and Kudo, A. 1973: Air pollution – especially in terms of oxidants and respiratory organs. *Intern. Med. (Japan)*, 32, 837–44.

Millar, I. B. and Cooney, P. A. 1982: Urban lead – a study of environmental lead and its significance to school children in the vicinity of a major trunk road. *Atmospheric Environ.*, 16, 615–20.

Miller, J. M. 1984: Acid rain. *Weatherwise*, 37, 222–51.

Miller, P. R., Parmeter, J. R., Flick, B. H. and Martinez, C. W. 1969: Ozone damage response of ponderosa pine seedlings. *J. Air Pollut. Control Ass.*, 19, 435–8.

Mills, E. S. and White, L. 1978: Government policies toward automotive emissions control. In A. F. Friedlaender (ed.), *Approaches to Controlling Air Pollution*, Cambridge, Mass.: MIT Press, 348–421.

Milne, A. 1979: *Noise Pollution: impact and countermeasures*. Newton Abbot: David & Charles.

Molina, M. J. and Rowland, F. S. 1974: Stratospheric sink for chlorofluoro-methanes: chlorine atom catalysed destruction of ozone. *Nature*, 249, 810–12.

Moncrieff, J. B. 1985: *The Identification and Assessment of Areas in Scotland which are 'At Risk' of Breaching EC Directive Limit Values for Smoke and Sulphur Dioxide*. Stevenage: Warrent Spring Laboratory, Report LR 531 (AP)M.

Morgan-Huws, D. I. and Haynes, F. N. 1973: Distribution of some epiphytic lichens around an oil refinery at Fawley, Hampshire. In B. W. Ferry, M. S. Baddeley and D. L. Hawksworth (eds), *Air Pollution and Lichens*, London: Athlone Press, 89–108.

Mote, V. L. 1974: Air pollution in the USSR. In I. Volgyes (ed.), *Environmental Deterioration in the Soviet Union and Eastern Europe*, New York: Praeger, 37–54.

Mulholland, K. A. and Attenborough, K. 1981: *Noise Abatement and Control*. London: Construction Press.

Munn, R. E. 1981: *The Design of Air Quality Monitoring Networks*. London: Macmillan.

Murozumi, M., Chow, T. J. and Patterson, C. C. 1969: Chemical concentration of pollutant lead aerosols, terrestrial dusts and sea salts in Greenland and Antarctic snow strata. *Geochem. Cosmochin. Acta*, 33, 1247–94.

Musk, L. F. 1982: The local fog hazard as a factor in planning new roads and motorways. *Environ. Education and Information*, 2, 119–29.

Naumann, P. J. 1973: Smoking and air pollution standards. *Science*, 182, 334–6.

Needleman, H. L. and Shapiro, J. M. 1974: Dentine lead levels in asymptomatic Philadelphian school children: subclinical exposure in high and low risk groups. *Environ. Health Perspect.*, Exp. Issue No. 7, 27–33.

Needleman, H. L., Gunnoe, C. E., Leviton, A., Reed, R., Peresie, H., Maher, C. and Barrett, P. 1979: Deficits in psychologic and classroom performance of children with elevated lead levels. *New England J. Med.*, 300, 689–95.

de Nevers, N. H. 1981: Measuring and managing pollutants. *Environment*, 23, 25–35.

de Nevers, N. H., Neligan, R. E. and Slater, H. H. 1977: Air quality management, pollution control strategies, modeling, and evaluation. In A. C. Stern (ed.), *Air Pollution. Vol. 5*, Air Quality Management, 3–40.

Newell, R. E. 1970: Water vapour pollution in the stratosphere by the Supersonic Transporter. *Nature*, 226, 70–1.

Nielding, von. G. and Wagner, H. M. 1979: Effects of nitrogen dioxide on chronic bronchitis. *Environ. Health Perspect.*, 29, 137–42.

Nielding, von. G., Wagner, H. M., Lollgen, H. and Krekeler, H. 1977: Acute effects of ozone function in man (in German). *VDI-Beriche*, 270, 123–9.

Nikolaou, K., Masclet, P. and Mouvier, G. 1984: Sources and chemical reactivity of polynuclear aromatic hydrocarbons in the atmosphere – a critical review. *Science of the Total Environ.*, 32, 103–32.

Noble, A. G. 1980: Noise pollution in selected Chinese and American cities. *Geojournal*, 4, 573–5.

Noble, A. G. and Harnapp, V. R. 1981: Towards a model for monitoring community noice. In L. J. C. Ma and A. G. Noble (eds), *The Environment: Chinese and American views*, New York: Methuen, 275–85.

Odén, S. 1968: The acidification of air and precipitation and its consequences on the natural environment (in Swedish). *Swedish Nat. Sci. Res. Council, Ecology Committee, Bull.* 1, 1–86.

Odén, S. 1971: Nederbordens forsurning-ett generellt hot mot ekosystemem. In I. Mysterud (ed.), *Forurensning og Biologisk Miljovern*. Universitetsforlaget, Oslov, 63–98.

OECD (Organization for Economic Co-operation and Development) 1977: *The OECD Programme on Long Range Transport of Air Pollutants: measurements and findings*. Paris: OECD.

OECD 1980a: *Environment Policies for the 1980s*. Paris: OECD.

OECD 1980b: *Noise Abatement Policies*. Paris: OECD.

OECD 1980c: *Pollution Charges in Practice*. Paris: OECD.

O'Hare, G. P. 1973: Lichens as techniques of pollution assessment. *Area*, 5, 223–9.

Ohmori, K. 1974: The effect of photochemical smog on humans, especially biological effects by low concentrations of ozone (in Japanese). *Environ. Pollut. Control*, 10, 1042–46.

Oke, T. R. 1973: City size and the urban heat island. *Atmospheric Environ.*, 7, 769–79.

Orehek, J., Massari, J. P., Gayard, P., Grimaud, C. and Charpin, J. 1976: Effect of short-term, low-level nitrogen dioxide exposure on bronchial sensitivity of asthmatic patients. *J. Clin. Invest.*, 57, 301–7.

O'Riordan, T. 1976: *Environmentalism*. London: Pion.

Ott, W. R. 1977: Development of criteria for siting air monitoring stations. *J. Air Pollut. Control Ass.*, 27, 543–47.

Ottar, B. 1981: The transfer of airborne pollutants to the Arctic region. *Atmospheric Environ.*, 15, 1439–45.

Overrein, L. N., Seip, H. M. and Tollan, A. 1980: *Acid Precipitation – Effects on Forest and Fish. Final Report of the SNSF Project 1972–1980*. Oslo, Norway.

Padgett, J. and Richmond, H. 1983: The process of establishing and revising National Ambient Air Quality Standards. *J. Air Pollut. Control Ass.*, 33, 13–16.

Padmanabhamurty, B. and Hirt, M. S. 1974: The Toronto Heat island and pollution distribution. *Water, Air and Soil Pollut.*, 3, 81–9.

Panofsky, H. A. 1978: Earth's endangered ozone. *Environment*, 20, 16–20, 40.

Parkinson, C. L. and Kellogg, W. W. 1979: Arctic sea ice decay simulated for a carbon dioxide-induced temperature rise. *Climatic Change*, 2, 149–62.

Parlour, J. W. 1980: The mass media and environmental issues: a theoretical analysis. *Intern. J. Environ. Studies*, 15, 109–21.

Parlour, J. W. and Schatzow, S. 1978: The mass media and public concern for environmental problems in Canada 1960–72. *Intern. J. Environ. Studies*, 13, 9–17.

Pashel, G. E. and Egner, D. R. 1981: A comparison of ambient suspended particulate matter concentrations as measured by the British Smoke Sampler and the High Volume Sampler at 16 sites in the United States. *Atmospheric Environ.*, 15, 919–27.

Patrick, R., Binetti, V. P. and Halterman, S. G. 1981: Acid lakes from natural and anthropogenic causes. *Science*, 211, 446–8.

Pengelly, L. D., Kerigan, A. T., Goldsmith, C. H. and Inman, E. M. 1984: The Hamilton Study: distribution of factors compounding the relationship between air quality and respiratory health. *J. Air Pollut. Control Ass.*, 34, 1039–43.

Pentreath, R. J. 1980: *Nuclear Power, Man and The Environment*. Wykeham London: Taylor & Francis, Wykeham Science Series No. 51.

Pepper, D. M. 1984: *The Roots of Modern Environmentalism*. London: Croom Helm.

Perry, A. H. and Walker, J. M. 1977: *The Ocean-Atmosphere System*. London: Longman.

Persson, G. 1977: Organization and operation of national air pollution control programs. In A. C. Stern (ed.) *Air Pollution. Vol. 5*, Air Quality Management, 381–413.

Peuschel, S. M. Kopito, L. and Schwachman, H. 1972: A screening and follow up study of children with an increased lead burden. *J. Amer. Med. Assoc.*, 333, 462–6.

Pollack, R. I. 1975: *Studies of Pollutant Concentration Frequency Distributions*. Washington, D. C.: US Environmental Protection Agency, EPA-650/4-7-004.

Pryde, P. R. 1972: *Conservation in the Soviet Union*. Cambridge: Cambridge University Press.

Pryde, P. R. 1983: The 'decade of the environment' in the USSR. *Science*, 220, 274–9.

Purdom, P. W. 1980: *Environmental Health*, 2nd. edition. New York: Academic Press.

Qu, G. 1982: Environmental protection in China. *Environ. Conservation*, 9, 31–3.

Qu, G. and Li, J. 1984: Environmental management. In Qu Geping and Woyen Lee (eds), *Managing the Environment in China*, Dublin: Tycooly International, 1–29.

Rabinowitz, M. and Needleman, H. L. 1983: Petrol lead sales and umbilical cord blood lead levels in Boston, Massachusetts. *Lancet*, 1, 63.

Rahn, K. A. 1981: Relative importances of North America and Eurasia as sources of arctic aerosol. *Atmospheric Environ.*, 15, 1447–55.

Ramanathan, V. 1975: Greenhouse effect due to chlorofluorocarbons: climatic implications. *Science*, 190, 50–2.

Ratcliffe, R. A. S. 1973: Recent work on sea-surface temperature anomalies related to long-range forecasting. *Weather*, 28, 106–17.

Ratcliffe, R. A. S. and Murray, R. 1970: New lag associations between North Atlantic sea temperatures and European pressure applied to long-range weather forecasting. *Quart. J. Royal Met. Soc.*, 96, 226–46.

Reck, R. A. 1976: Stratospheric ozone effects on temperature. *Science*, 192, 557–9.

Record, F. A., Bubenick, D. V. and Kindya, R. J. 1982: *Acid Rain Information Book*. New Jersey: Noyes Data Corporation.

Reinsel, G. C. 1981: Analysis of total ozone data for the detection of recent trends and the effects of nuclear testing during the 1960s. *Geophys. Res. Letters*, 8, 1227–30.

Reinsel, G. C., Tiao, G. C., Wang, M. N., Lewis, R. and Nychka, D. 1981: Statistical analysis of stratospheric ozone date for the detection of trends. *Atmospheric Environ.*, 15, 1569–77.

Renberg, I. and Hellberg, T. 1982: The pH history of lakes in SW Sweden, as calculated from the subfossil diatom flora of the sediments. *Ambio*, 11, 30–3.

Revelle, R. 1982: Carbon dioxide and world climate. *Scientific American*, 247, 33–41.

Revelle, R. R. and Shapero, D. C. 1978: Energy and climate. *Environ. Conservation*, 5, 81–91.

Rombout, P. J. A. Lioy, P. J. and Goldstein, B. D. 1986: Rationale for an eight-hour ozone standard. *J. Air Pollut. Control Ass.*, 36, 913–7.

Rosenberg, N. J. 1981: The increasing carbon dioxide concentration in the atmosphere and its implication on agricultural productivity. I. Effects on photosynthesis, transpiration and water use efficiency. *Climatic Change*, 3, 265–79.

Rosenberg, N. J. 1982: The increasing carbon dioxide concentration in the atmosphere and its implication on agricultural productivity. II. Effects through carbon dioxide-induced climatic change. *Climatic Change*, 4, 239–54.

Rosencranz, A. 1981: Economic approaches to air pollution control. *Environment*, 23, 25–30.

Rosencranz, A. 1982: EEC moves to control sulphur dioxide pollution . . . at a snail's pace. *Ambio*, 11, 371–3.

Rosencranz, A. 1986: The acid rain controversy in Europe and North America: a political analysis. *Ambio*, 15, 47–51.

Rowland, F. S. 1988: Chlorofluorocarbons, stratospheric ozone, and the Antarctic 'ozone hole'. *Environmental Conservation*, 15, 101–15.

Samberg, A. and Elston, J. 1982: The proposed revision to the NAAQS for carbon monoxide and its effect on state designation. *J. Air Pollut. Control Ass.*, 32, 1228–30.

Sand, P. H. 1985: The Vienna Convention is adopted. *Environment*, 27, 18–20, 40–3.

Sandbach, F. 1980: *Environment, Ideology and Policy*. Oxford: Basil Blackwell.

Sandbach, F. 1982: *Principles of Pollution Control*. London: Longman.

Sawyer, R. F. and Pitz, W. J. 1983: *Assessment of the Impact of Light Duty Diesel Vehicles on Soiling in California. Report to the California Air Resources Board, 25 January 1983.*

Scarrow, H. A. 1972: The impact of British domestic air pollution legislation. *Brit. J. Polit. Science*, 2, 261–82.

Schimmel, H. 1978: Evidence for possible health effects of ambient air pollution from time series analysis: methodological questions and some new results based on New York City daily mortality, 1963–76. In *Proc. Sympos. Environ. Effects of Sulphur Oxides & Related Particulates, vol. 54*, New York Academy of Medicine, 1052–108.

Schimmel, H. and Murawski, T. J. 1976: The relation of air pollution to mortality. *J. Occup. Med.*, 18, 316–33.

Schofield, C. L. 1976: Acid precipitation: effects on fish. *Ambio*, 5, 228–30.

Schrenk, H. H., Heinmann, H., Clayton, G. D., Gafafer, W. M. and Wexler, H. 1949: Air pollution in Donora, Pa. Epidemiology of the unusual smog episode of October 1948. *Public Health Bulletin*, 306, Washington, D. C.: Public Health Service.

Schweitzer, G. E. 1983: Toxic chemicals: steps towards their evaluation and control. In D. A. Kay and H. K. Jacobson (eds), *Environmental Protection: the international dimension*, Allanheld: Osmun, 22–44.

SCOPE (Scientific Committee on Problems of the Environment) 1985: *Environmental Consequences of Nuclear War.* Chichester & New York: John Wiley & Sons, SCOPE 28, 2 volumes.

Sellers, L. and Jones, D. W. 1973: Environment and the mass media. *J. Environ. Education*, 5, 51–7.

Shaikh, R. A. and Nichols, J. K. 1984: The international management of chemicals. *Ambio*, 13, 88–91.

Shaver, C. L., Cass, G. R. and Druzik, J. R. 1983: Ozone and the deterioration of works of art. *Environ. Sci.Technology*, 17, 748–52.

Shaw, G. E. 1982: Evidence for a central Eurasian source area of Arctic haze in Alaska. *Nature*, 299, 815–18.

Shen, T. T. 1984: Environmental management in mainland China. *Environmentalist*, 4, 317–21.

Shreffler, J. H. 1979: Heat island convergence in St Louis during calm periods. *J. App. Meteor.*, 18, 1512–20.

Shy, C. M. and Finklea, J. F. 1973: Air pollution affects community health. *Environ. Sci. Technology*, 7, 204–8.

Shy, C. M. and Love, G. J. 1980: Recent evidence on the human health effects of nitrogen dioxide. In S. D. Lee (ed.), *Nitrogen Oxides and Their Effects on Health*, Ann Arbor: Ann Arbor Science, 291–305.

Siddiqi, T. A. and Zhang, C-X. 1984: Ambient air quality standards in China. *Environ. Management*, 8, 473–9.

Sidebottom, H. 1979: Impact of pollutants on the ozone layer. *In Proc. of Seminar on Air Pollution: impacts and control*, Dublin: National Board for Science & Technology, 52–60.

Sionit, N.,Strain, B. R., Hellmers, H. and Kramer, P. J. 1981: Effects of atmospheric carbon dioxide concentration and water stress on water relations of wheat. *Botanical Gazette*, 142, 191–6.

Skarby, L. and Sellden, G. 1984: The effects of ozone on crops and forests. *Ambio*, 13, 68–72.

Smil, V. 1980a: China's environment. *Current History*, 79, 14–18.

Smil, V. 1980b: Environmental degradation in China. *Asia Survey*, 20, 777–8.

Smil, V. 1984: *The Bad Earth: environmental degradation in China.* New York: Sharpe; and London: Zed Press.

Smith, F. B. and Hunt, R. D. 1978: Meteorological aspects of the transport of pollution over long distances. *Atmospheric Environ.*, 12, 461–77.

Smith, R. J. 1981a: EPA and industry pursue regulatory options. *Science*, 211, 796–8.

Smith, R. J. 1981b: The fight over clean air begins. *Science*, 211, 1328–30.

Speizer, F. E., Ferris Jr, B. Bishop, Y. M. M. and Spengler, J. 1980: Health effects of indoor nitrogen dioxide exposure: preliminary results. In S. D. Lee (ed.), *Nitrogen Oxides and Their Effects on Health*, Ann Arbor: Ann Arbor Science, 343–59.

Stafford, H. A. 1977: Environmental regulations and the location of US manufacturing: speculations. *Geoforum*, 8, 243–8.

Starkie, D. N. M. 1976: The spatial dimensions of pollution policy. In J. T. Coppock and W. R. D. Sewell (eds), *Spatial Dimensions of Public Policy*, Oxford: Pergamon, 148–63.

Stensland, G. J. and Semonin, R. G. 1982: Another interpretation of the pH trend in the United States. *Bull. Amer. Met. Soc.*, 63, 1277–84.

Stern, A. C. 1977: Prevention of Significant Deterioration: a critical review. *J. Air Pollut. Control Ass.*, 27, 440–53.

Stern, A. C. 1982: History of air pollution legislation in the United States. *J. Air Pollut. Control Ass.*, 32, 44–61.

Stoel, T. B. Jr. 1983 Fluorocarbons: mobilizing concern and action. In D. A. Kay and H. K. Jacobson (eds), *Environmental Protection. The international dimension*, Allanheld, Osmun, 45–74.

Stolarski, R. S. 1988: The Antarctic ozone hole. *Scientific American*, 258, 20–6.

Svensson, B. H. and Soderlund, R. 1975: *Nitrogen, Phosphorus and Sulphur – Global Cycles*. Orsundsbro: Royal Swedish Academy of Sciences, SCOPE Report 7.

Swannack-Nunn, S., Bowman, K. and Heffernan, P. 1979: *Environmental Protection in the People's Republic of China*. Washington, D. C.: National Council for US-China Trade.

Swedish Ministry of Agriculture 1982: *Acidification Today and Tomorrow*. (Report prepared for the 1982 Stockholm Conference on the Acidification of the Environment by the Environment '82 Committee) Swedish Ministry of Agriculture.

Tabershaw, I. R. and Gaffey, W. R. 1974: Mortality study of workers in the manufacture of vinyl chloride and its polymers. *J. Occup. Med.*, 16, 509–18.

Temple, P. J. and Taylor, O. C. 1983: World-wide ambient measurements of peroxyacetyl nitrate (PAN) and implications for plant injury. *Atmospheric Environ.*, 17, 1583–7.

Thomas, A. J. and Martin, J. M. 1986: First assessment of Chernobyl radioactive plume over Paris. *Nature*, 321, 817–19.

Thompson, G. 1978: *The Museum Environment*. London: Butterworths.

Thompson, S. L., Alexandrov, V. V., Stenchikov, G. L., Schneider, S. H., Covey, C. and Chervin, R. M. 1984: Global climatic consequences of nuclear war: simulations with three-dimensional models. *Ambio*, 13, 236–43.

Thornes, J. E. 1977: Ozone comes to London. *Progress in Physical Geography*, 1, 506–17.

Timberlake, L. 1981: Poland – the most polluted country in the world? *New Scientist*, 92, 248–50.

Tomlinson, G. H. and Silversides, C. R. 1982: *Acid Deposition and Forest Damage – The European Linkage*. Montreal: Domtar, Inc.

Tout, D. G. 1978: Mortality in the June–July 1976 hot spell. *Weather*, 33, 221–6.

Tout, D. G. 1979: The improvement in winter sunshine totals in city centres. *Weather*, 34, 67–71.

Trijonis, J. 1979: Visibility in the Southwest – an exploration of the historical data base. *Atmosphere Environ.*, 13, 833–43.

Trofimenko, S. 1983: The Soviet Union tightens up its safety requirements for nuclear power plants. *Ambio*, 12, 360.

Turco, R. P., Toon, O. B., Ackerman, T. P., Pollack, J. B. and Sagan, C. 1983: Nuclear winter: global consequences of multiple nuclear explosions. *Science*, 222, 1283–92.

UK Department of Environment 1979a: Chlorofluorocarbons and their effect on stratospheric ozone (second report). *Pollution Paper, Central Directorate on Environ. Pollution,* London, HMSO, 15.

UK Department of Environment 1979b: The United Kingdom environment 1979: progress of pollution control. *Pollution Paper, Central Directorate on Environ. Pollution,* London, HMSO, 16.

UK Department of Environment 1983: Lead in the environment: the government response to the ninth report of the Royal Commission on Environmental Pollution. *Pollution Paper, Central Directorate on Environ. Pollution,* London, HMSO, 19.

UK Department of Environment 1984a: Controlling pollution: principles and prospects. The Government's response to the Tenth Report of the Royal Commission on Environmental Pollution. *Pollution Paper, Central Directorate on Environ. Pollution,* London, HMSO, 22.

UK Department of Environment 1984b: *Digest of Environmental Pollution and Water Statistics No. 6 (1983)*. London: HMSO.

UK Department of Environment 1985: *Digest of Environmental Protection and Water Statistics, No. 7 (1984)*. London: HMSO.

UK Department of Environment 1986: *Digest of Environmental Protection and Water Statistics, No. 8 (1985)*. London: HMSO.

UK Department of Health and Social Security 1980: *Lead and Health: the Report of a DHSS Working Party on lead in the environment*. London: HMSO.

UK Department of Trade and Industry 1983: Air Pollution. *Technology and the Environ.*, 12(3), 51–6.

UK Department of Trade and Industry 1984: Air Pollution. *Technology and the Environ.*, 12(4), 35–45.

UK Greater London Council 1979: *Public Services and Safety Committee Report, 15 February 1979*. London: GLC.

UK HMSO (Her Majesty's Stationery Office) 1981: *Coal and the Environment*. London: HMSO.

UK House of Commons Environment Committee 1984: *Acid Rain*. Session 1983–84. London: HMSO.

UK House of Lords Select Committee on the European Communities 1984: *Air Pollution*. 22nd Report. Session 1983–84. London: HMSO.

UK Open University 1975: *Air Pollution Control*. Milton Keynes: Open University, Environmental Control and Public Health Unit 15.

UK Royal Commission on Environmental Pollution 1971: *First Report*. London: HMSO.

UK Royal Commission on Environmental Pollution 1974: *Fourth Report: Pollution Control: progress and problems*. London: HMSO.

UK Royal Commission on Environmental Pollution 1976: *Fifth Report: Air Pollution Control – an integrated approach*. London: HMSO.

UK Royal Commission on Environmental Pollution 1983: *Ninth Report: Lead in the Environment*. London: HMSO.

UK Royal Commission on Environmental Pollution 1984: *Tenth Report: Tackling Pollution – Experience and Prospects*. London: HMSO.

Ulrich, B. and Pankrath, J. 1983: *Effects of Accumulation of Air Pollutants in Forest Ecosystems*. Boston: Reidel.

Ulrich, B., Mayer, R. and Khanna, P. K. 1980: Chemical changes due to acid precipitation in a löss derived soil in central Europe. *Soil Science*, 130, 193–9.

United Nations Economic Commission for Europe 1985: *Air Pollution Across National Boundaries*. New York: United Nations, Air Pollution Studies 2.

UNSCEAR (United Nations Committee on the Effects of Atomic Radiation) 1977: *Sources and Effects of Ionising Radiation*. New York: UNSCEAR.

UNSCEAR (United Nations Committee on the Effects of Atomic Radiation) 1982: *Ionising Radiation: Sources and Biological Effects*. New York: UNSCEAR.

Unsworth, M. H., Shakespeare, N. W., Milner, A. E. and Ganendra, T. S. 1979: The frequency of fog in the Midlands of England. *Weather*, 34, 72–7.

Urquhart, J. 1983: Polonium: Windscale's most lethal legacy. *New Scientist*, 97, 873–5.

US Board on Toxicology and Environmental Hazards 1979: *Odors from Stationary and Mobile Sources*. Washington, D.C.: National Academy of Sciences.

US Board on Toxicology and Environmental Hazards 1980: Odors from stationary and mobile sources. *J. Air Pollut. Control Ass.*, 30, 13–16.

US EPA (Environmental Protection Agency) 1983a: *Can We Delay a Greenhouse Warming?* Washington, D.C.: US Environmental Protection Agency, US Government Printing Office.

US EPA 1983b: *National Air Quality and Emissions Trends Report, 1981*. Research Triangle, North Carolina: US Environmental Protection Agency, EPA-450/4-83-011.

US EPA 1986: *National Air Quality and Emissions Trends Report, 1984*. Research Triangle Park, North Carolina: UK Environmental Protection Agency, EPA-450/4-86-001.

US General Accounting Office 1979: *Air Quality: do we really know what it is?* Washington D.C.: General Accounting Office, Report CED 79–84.

US National Clean Air Coalition 1981: Cleaning up the Clean Air Act. *Environment*, 23, 16, 20, 42–4.

US National Commission on Air Quality 1981: *To Breathe Clean Air*, Washington, D.C.

US National Environmental Development Association 1981: Cleaning up the Clean Air Act. *Environment*, 23, 17–20.

US National Research Council 1975a: *Environmental Impact of Stratospheric Flight: biological and climatic effects of aircraft emissions in the stratosphere.* Washington, D.C.: National Academy Press.

US National Research Council 1975b: *Long-term Worldwide Effects of Multiple-Nuclear Weapon Detonations.* Washington, D.C.: National Academy Press.

US National Research Council 1977a: *Carbon Monoxide.* Washington, D.C.: National Academy Press.

US National Research Council 1977b: *Energy and Climate.* Washington, D.C.: National Academy Press, Studies in Geophysics. Geophysics Study Committee.

US National Research Council 1979: *Carbon Dioxide and Climate: a scientific assessment.* Washington, D.C.: National Academy Press, Ad Hoc Study Group.

US National Research Council 1982a: *Causes and Effects of Stratospheric Ozone Reduction: an update.* Washington, D.C.: National Academy Press.

US National Research Council 1982b: *Carbon Dioxide and Climate: a second assessment.* Washington, D.C.: National Academy Press, Carbon Dioxide/Climate Review Panel.

US National Research Council 1983a: *Acid Deposition. Atmospheric Processes in Eastern North America.* Washington, D.C.: National Academy Press.

US National Research Council 1983b: *Changing Climate.* Washington, D.C.: National Academy Press, Carbon Dioxide Assessment Committee.

US National Research Council 1984: *Causes and Effects of Changes in Stratospheric Ozone: update 1983.* Washington, D.C.: National Academy Press.

US NUREG-0558 (Nuclear Regulatory Commission) 1979: *Population Dose and Health Impact of the Accident at Three Mile Island Nuclear Station.* Washington, D.C.: US Nuclear Regulatory Commission.

US NUREG-0600 1979: *Investigation into the March 28, 1979 Three Mile Island Accident.* Washington, D.C.: U.S. Nuclear Regulatory Commission.

US SCAQMD (South Coast Air Quality Management District) 1985: *Air Quality Trends in the South Coast Air Basin 1975–1984.* El Monte, California: SCAQMD.

US SCAQMD 1986: *Air Quality Data 1985.* El Monte, California: SCAQMD.

Waggoner, P. E. 1984: Agriculture and carbon dioxide. *Amer. Scientist*, 72, 179–84.

Walker, C. 1975: *Environmental Pollution by Chemicals.* 2nd edition. London: Hutchinson.

Walker, H. M. 1985: Ten-year ozone trends in California and Texas. *J. Air Pollut. Control Ass.*, 35, 903–12.

Wall, G. 1973: Public response to air pollution in South Yorkshire, England. *Environment and Behavior*, 5, 219–48.

Wall, G. 1974a: Complaints concerning air pollution in Sheffield. *Area*, 6, 3–8.

Wall, G. 1974b: Public response to air pollution in Sheffield, England. *Intern. J. Environ. Studies*, 5, 259–70.

Wall, G. 1976a: Air pollution. *Progress in Geography*, 8, 96–131.

Wall, G. 1976b: National coping styles: policies to combat environmental problems. *Intern. J. Environ. Studies*, 9, 239–45.

Wallace, L. A. and Ziegenfus, R. C. 1985: Comparison of carboxyhemoglobin concentrations in adult non-smokers with ambient carbon monoxide levels. *J. Air Pollut. Control Ass.*, 35, 944–9.

Walsh, P. J., Dudney, C. S. and Copenhaver, E. M. 1983: *Indoor Air Quality*. Boca Raton, Florida: CRC Press, Inc.

Ward, M. A. 1981: The Clean Air Act controversy: Congress confronts the issues. *Environment*, 23, 6–15.

Wark, K. and Warner, C. F. 1981: *Air Pollution: its origins and control*. 2nd edition. New York: Harper & Row.

Warmbt, W. 1979: Ergebnisse langjähriger messungen des bodennahen ozons in der DDR. *Zeitschift für Meteorologie*, 29, 24–31.

Warner, Sir F. 1979: Sources and extent of pollution. *Proc. Roy. Soc. London, B*, 205, 5–15.

Weber, E. 1982: *Air Pollution: assessment methodology and modeling. Vol. 2.* New York: Plenum Pres.

Weil, M. 1981: Cleaning up China's environment. *China Business Review*, 8, 50–4.

Wetstone, G. S. and Foster, S. A. 1983: Acid precipitation: what is it doing to our forests? *Environment*, 25, 10–12, 38–40.

Wetstone, G. S. and Rosencranz, A. 1983: *Acid Rain in Europe and North America: national responses to an international problem*. Washington, D.C.: Environmental Law Institute.

White, W. H., Anderson, J. A., Blumenthal, D. L., Husar, D. L., Gillani, R. B. Husar, J. D. and Wilson, W. E. Jr. 1976: Formation and transport of secondary air pollutants: ozone and aerosols in the St Louis urban plume. *Science*, 194, 187–9.

Whitten, R. C., Borucki, W. J. and Turco, R. P. 1975: Possible ozone depletions following nuclear explosions. *Nature*, 257, 38–9.

Wigley, T. M. L., Jones, P. D. and Kelly, P. M. 1980: Scenario for a warm, high-carbon dioxide world. *Nature*, 283, 17–21.

Wilde, P. J. 1978: EEC approach to air pollution control – the government viewpoint. *Clean Air*, 8, 14–22.

Williams, J. 1978: The effects of climate on energy policy. *Electronics and Power*, 24, 261–8.

Wilson, A. T. 1978: Pioneer agriculture explosion and carbon dioxide levels in the atmosphere. *Nature*, 273, 40–1.

Wilson, D. 1983: *The Lead Scandal*. London: Heinemann.

Index